IOT FOR SMART OPERATIONS IN THE OIL AND GAS INDUSTRY

IOT FOR SMART OPERATIONS IN THE OIL AND GAS INDUSTRY
From Upstream to Downstream

RAZIN FARHAN HUSSAIN
School of Computing and Informatics
University of Louisiana at Lafayette
Lafayette, LA, United States

ALI MOKHTARI
School of Computing and Informatics
University of Louisiana at Lafayette
Lafayette, LA, United States

ALI GHALAMBOR, PH.D., P.E.
International Consultant
Laguna Niguel, CA, United States

MOHSEN AMINI SALEHI
School of Computing and Informatics
University of Louisiana at Lafayette
Lafayette, LA, United States

Gulf Professional Publishing
An imprint of Elsevier

For information on all Gulf Professional Publishing publications
visit our website at https://www.elsevier.com/books-and-journals

Publisher: Charlotte Cockle
Senior Acquisitions Editor: Katie Hammon
Editorial Project Manager: Howie M. De Ramos
Production Project Manager: Prasanna Kalyanaraman
Designer: Mark Rogers

Typeset by VTeX

To my parents, H.K.M. Altaf Hussain and Fariha Akhter, who taught me perseverance is the key to success

Razin Farhan Hussain

To my daughters, Mana and Niki Amini Salehi, without whom we could finish authoring this book much earlier! ;-)

Contents

Biography

Razin Farhan Hussain

Razin Farhan Hussain is a software quality engineer at TryCycle Data Systems Inc., Canada. He is also a Ph.D. candidate in the computer science department at the University of Louisiana at Lafayette. Besides his regular job, Razin works as a research assistant at the computer science department's High Performance Cloud Computing (HPCC) Lab. His research interest includes efficient utilization of fog computing for Industry 4.0 applications focusing on Deep Neural Network models. He earned an M.Sc. in computer science from the University of Louisiana at Lafayette, and a Bachelor of Science in computer science and engineering from the Military Institute of Science and Technology at the Bangladesh University of Professionals.

Ali Mokhtari

Ali Mokhtari is currently a Ph.D. candidate in computer science at the University of Louisiana at Lafayette. Ali works as a research assistant at High-Performance Cloud Computing (HPCC) Lab in the computer science department. His research interest includes using reinforcement learning for efficient resource allocation in heterogeneous edge computing systems. He earned an M.Sc. in aerospace engineering from the Sharif University of Technology in Iran and a B.Sc. in mechanical engineering from Shiraz University in Iran.

Ali Ghalambor (Ph.D., P.E.)

Prof. Ali Ghalambor, who is currently an international consultant, has 45 years of industrial and academic experience in the petroleum and mineral industries. He held engineering and supervisory positions at Marlin Drilling, Tenneco Oil, Amerada Hess Corporation, and Occidental Research Corporation. He previously served as the American Petroleum Institute (API) Endowed Professor, Head of the Petroleum Engineering Department, and Director of the Energy Institute at the University of Louisiana at Lafayette. He was also the Technical Director and a Program Manager at the Qatar National Research Fund of Qatar Foundation. Dr. Ghalambor has performed consulting and training services in 45 countries.

He has authored or coauthored 16 books and manuals and more than 250 technical papers. His technical contributions on fundamental and applied research on formation damage control, well drilling, well completions, and production operations are internationally recognized.

He has received many of the Society of Petroleum Engineers (SPE) and the API prestigious awards, including the Production and Operations Award, the Distinguished Achievement Award for Petroleum Engineering Faculty, the Distinguished Member Award, and the DeGolyer Distinguished Service Award. Furthermore, he is the recipient of the Robert Earll McConnell Award, which is a joint SPE and AIME Award that recognizes beneficial service to mankind by engineers through significant contributions that advance a nation's standard of living or replenish its natural resource base. He served as a commissioner on the Engineering Accreditation commission of ABET and was Director of the Central and Southeastern North America Region on the SPE International Board of Directors and is the founding chairman of the SPE International Conference and Exhibition on Formation Damage Control. Dr. Ghalambor holds BS and MS degrees in petroleum engineering from the University of Southwestern Louisiana and a PhD from Virginia Polytechnic Institute and State University. He is a Registered Professional Engineer in the U.S. State of Texas and the District of Columbia and an elected member of the Russian Academy of Natural Sciences and recipient of its Nobel Laureate Physicist Kapitsa Gold Medal.

Mohsen Amini Salehi (Ph.D.)

Dr. Mohsen Amini Salehi is an Associate Professor and Francis Patrick Clark/BORSF Endowed Professorship holder at the School of Computing and Informatics (CMIX), University of Louisiana Lafayette, USA. He is the director of High Performance and Cloud Computing (HPCC) Laboratory where several graduate and undergraduate students research on various aspects of Distributed and Cloud computing. Dr. Amini is an NSF CAREER Awardee and, so far, he has had 11 research projects funded by National Science Foundation (NSF) and Board of Regents of Louisiana (totaling $2,722,322). He has also received five awards and certificates from University of Louisiana at Lafayette in recognition of his innovative research.

Dr. Amini has received his Ph.D. in Computing and Information Systems from Melbourne University, Australia, in 2012, under supervision of Professor Rajkumar Buyya. He was awarded postdoctoral fellowship at Colorado State University (2012–2013), and University of Miami (2013–2014), before joining UL Lafayette as a faculty member in 2014. He has completed his Master's and Bachelor's degree in Computer Software Engineering from Ferdowsi University of Mashhad and Azad University of Mashhad in 2006 and 2003, respectively.

Dr. Amini has been an active researcher in Distributed and Cloud computing research areas since 2004. He has received several awards in recognition of his research, including the "Best Intern Award" from Infosys Ltd., in 2012, and the IEEE award for "Applied Research", in 2009. His paper was nominated for the "Best Paper Award" in 33rd International Parallel and Distributed Processing Symposium (IPDPS '19), Rio de Janeiro, Brazil.

Dr. Amini has 64 publications that have received more than 1200 citations so far. He has filed 4 U.S. patents as a result of his research in Distributed and Cloud computing. Dr. Amini has been active in the professional community. He has served as the technical program committee and organizer for several premier conferences, including SuperComputing 2022 (SC '22), IEEE International Cloud Computing (CLOUD '22), IEEE/ACM International Symposium in Cluster, Cloud, and Grid Computing (CCGrid '20 and '21), International Green and Sustainable Computing Conference (IGCC '18) among several others. He constantly reviews papers for top-notch journals such as IEEE Transactions on Parallel and Distributed Systems (TPDS), IEEE Transactions on Cloud Computing (TCC), Journal of Parallel and Distributed Systems (JPDC), and Future Generation Computing System Journal (FGCS) and constantly serves in NSF panels. His research interests are in building smart systems across edge-to-cloud continuum, virtualization, resource allocation, energy-efficiency, heterogeneity, and trustworthy in Distributed, Edge, and Cloud computing systems.

Preface

Industrial systems, particularly Oil and Gas, are rapidly shifting from human-controlled processes towards closed-loop control systems that leverage massive sensing and computing infrastructure to manage their operations autonomously. This paradigm shift is a key to enable emerging data-intensive and delay-sensitive Industry 4.0 applications, particularly at remote sites, *e.g.*, offshore Oil and Gas (O&G) fields, where the access to computing infrastructure is limited and human resources are scarce. Realizing these systems demands interdisciplinary research and training at the intersection of Industry 4.0 in O&G, modern computing infrastructure (such as a Edge and Cloud), and Machine Learning to nurture the next generations of engineers and scientists who can handle the emerging problems of these systems.

Accordingly, this book aims at introducing the opportunities and solutions to develop a smart and robust O&G industry based on the principles of the Industry 4.0 paradigm. As a result, the book will enable the researchers and practitioners in the IT industry to play a key role in making the O&G industry safer, more sustainable, greener, automated, and eventually more cost-efficient. For that purpose, the book explores the methods of employing computing technologies throughout the Oil and Gas industry—from upstream to midstream, and downstream sectors. It elaborates on how the synergy between state-of-the-art computing platforms, such as Internet of Things (IoT) and cloud computing and Machine Learning methods, can be harnessed to serve the purpose of a more efficient O&G industry. Our approach in this book is to explore the operations performed in each sector of the O&G industry and then, introduce the computing platforms and methods that can potentially enhance the operation. We note that smart O&G is a double-edge sword and not everything about the smart solutions is bright. There are also dark sides, in terms of threats and side-effects. This book pays a particular attention to these dark sides and dedicates a chapter to understand and provide solutions for them.

In sum, the key elements of the book are as follows: (A) Well-defined and comprehensive taxonomy of various sectors of O&G industry; (B) Describing solutions based on IoT, Cloud, and Machine Learning, to help the O&G industry transitioning to the efficient and safe Industry 4.0 paradigm; (C) An end-to-end architecture that encompasses a range of operations from data collection via sensors to process them in real-time via Machine

Learning methods in a continuum of edge-to-cloud systems; (D) Threats and side-effects of smart O&G and solutions to overcome them; and (E) Case-studies of oil-spill detection and other critical scenarios using deep neural network models on a federation of Edge and Cloud systems.

This book targets audience in both academia and industry. The most relevant target audience is petroleum engineers and Information Technology (IT) associates aiming at acquiring the knowledge to conduct projects to improve the efficiency of the O&G industry via state-of-the-art computing technologies and methods. It is this gap between computer scientists and petroleum engineers that the book attempts to bridge. In particular, computer researchers and practitioners desire resources that offer them domain knowledge in the petroleum industry, so that they can develop and tailor solutions for the O&G industry. However, the petroleum industry is highly complex and there is scarcity of comprehensive petroleum knowledge. As such, this book provides a systematic and categorical explanation of the complex O&G extraction process along with possible solutions using cutting-edge technologies that can benefit O&G strategists and decision makers. This book can be a primary source for the students in the emerging interdisciplinary area of *energy informatics* that focuses on the applications of advanced IT technology in energy production and distribution. It can be also used as the source for the interdisciplinary courses, such as Industrial IoT, that is offered to graduate and senior-level undergraduate students in petroleum, chemical, mechanical, and other engineering disciplines.

This book is a collaborative and interdisciplinary effort between Dr. Amini and his research team (Razin Farhan Hossain and Ali Mokhtari) at the High Performance Cloud Computing (HPCC) research group (http:// hpcclab.org) at University of Louisiana Lafayette, and Dr. Ghalambor who is an experienced research scientist and practitioner in the O&G industry. Our team has accumulated more than four years of knowledge and experience in smart O&G industry that provides us with the vision needed to author this book.

Mohsen Amini Salehi
Lafayette, LA, United States
April 2022

CHAPTER 1

Introduction to smart O&G industry
Overview of smart O&G industry

1.1 Challenges of the O&G industry

With the advancement in computing, communication, and IoT, the O&G industry is getting more competent and well-equipped with cyber-physical devices to perform routines and safety-related operations in a robust manner [1–3]. These cyber-physical systems are composed of sensors, gauges, and actuators that generate a considerable amount of data every day. The data needs to be processed and analyzed to respond to particular recipients, such as actuators, motors, gauges, and pumps, to complete any particular operation. The operation or functionality could be delay-tolerant or have a real-time nature [4]. In either case, computing and processing have become an indispensable part of the oil and gas industry.

While many of the above-mentioned characteristics exist in other industrial systems too, the O&G industry faces particular challenges that need a specific attention. One challenge is due to operating in remote (often offshore) areas, where the access to cloud and other back-end computing services is limited, and human resources are scarce too. Under these circumstances, realizing the idea of smart O&G mandates dealing with obstacles in diverse areas, such as real-time data collection and processing, low latency (submilliseconds) wireless data communications, robustness against uncertainties and accidents, and smart decision making, just to name a few. Another challenge is the scope of O&G industry that encompasses different sectors and specializations. Providing unified solutions across these sectors and specialized operating teams/companies is difficult, is not impossible. As such, there is a growing demand for interdisciplinary research and development at the intersection of computing, O&G, and other specializations related to these areas to nurture the future generations of engineers and scientists with a domain knowledge in developing smart O&G solutions.

IoT for Smart Operations in the Oil and Gas Industry
https://doi.org/10.1016/B978-0-32-391151-1.00010-1

Fig. 1.1 shows a bird-eye view of the operational workflow of the O&G industry. Its challenges can be divided into three main sectors: upstream, midstream, and downstream. Although this categorization is provided from the O&G industry perspective, we consider it as the reference model for this book, and its components are explored across different chapters.

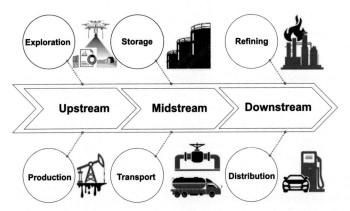

Figure 1.1 Workflow of oil and gas industry from upstream to downstream. The major operations are demonstrated related to corresponding sectors.

To the best of our knowledge, this is the very first comprehensive survey of computing aspects in the smart O&G domain. Nowadays, O&G industries are increasingly digitalized with cutting-edge hardware and software technologies to perform various automated activities (*e.g.*, for exploration surveys, drilling operations, pipeline monitoring, etc.). It is a challenge for petroleum experts to understand, explore, and implement the state-of-the-art computing technologies efficiently across the upstream, midstream, and downstream sectors. That is why, many advanced computing solutions are either not effectively utilized or not even well thought of across various O&G sectors. The same goes for computer experts who generally lack the domain knowledge of petroleum processes to develop effective software and hardware solutions. Due to this gap, new concepts and technologies are poorly implemented in this domain and the existing technological potentials have remained unleashed. *It is this knowledge gap between the petroleum industry and the information technology that this book aims to fill*, so that they become synergistic and together push the boundaries in the emerging interdisciplinary area of *energy informatics*.

1.2 Objectives of the smart O&G industry

The primary goal of the O&G industry is to maximize the production and minimize the side-effects. Their specific list of objectives is as follows:

(a) Maximizing the production

(b) Minimizing the incurred cost

(c) Safety of the operations for both the general public and the workers

(d) Minimizing the environmental impacts

To accomplish the above-mentioned objectives, the O&G industry has to perform activities in the following directions:

- Developing management and technical strategies specific to each sector.
- Controlling the operation of the equipment in each sector.
- Monitoring the operations of each sector.

The fundamental part of these activities is the control and monitoring systems. The monitoring system surveils any particular area or status of an equipment. The control system processes the observed area or status via sensors and take actions through actuators. Therefore, monitoring and controlling systems are heavily technology-dependent and work hand in hand for the smooth operation of the system. Monitoring and controlling systems together form a field of technology called *Cyber-Physical Systems* (CPS). A cyber-physical system deals with two aspects: the cyber aspect that refers to the sensors/actuators and the computing systems; and the physical aspect that deals with the equipment, and the human who work with them.

In sum, the smart O&G industry is considered as a large-scale and complex CPS system that studying it requires deep understanding of all the computing and physical aspects. In the next sections of this chapter, we elaborate on these aspects and then, we explain how different chapters of this book cover them.

1.3 Smart O&G: computing and middleware aspects

1.3.1 Landscape of computing infrastructure for O&G industry

Modern computing systems, such as cloud and edge, enable the smooth operation of different fault-intolerant processes across different sectors of the O&G industry. As a cyber-physical system, the computing system of the O&G industry is composed of the following components:

- *Sensors:* Numerous sensors of different types (*e.g.*, to gauge pressure, emission of toxic gases, security cameras, etc.) continuously procure multi-modal data in the form of signal, text, images, video, and audio. The data is stored or communicated for offline or online processing to monitor the operation of the oil field or to make management decisions.

- *Computer networks:* In a smart oil field, short- and long-range wireless and wired computer networks (*e.g.*, Bluetooth, CBRS, satellite, etc.) have to be configured for low-latency and high data-rate communication of devices (*e.g.*, sensors, servers, and actuators) both for onsite and offsite communication.

- *Computing systems and middleware:* All the collected data have to be eventually processed to be useful. That is why, in the back-end, smart oil fields are reliant on different forms of computing systems (*e.g.*, HPC, cloud, fog, edge, and real-time systems) to perform batch or online data processing for purposes like monitoring, visualization, and human-based or automatic decision making.

- *Data processing and software technologies:* The rule of thumb in a smart oil field is that "the more data can be processed, the more informed decisions can be made". The large amount of multi-modal data (text, images, video, and signals) continuously generated in a smart oil field form what is known as *big data*. Such diverse formats of data have to be processed using various algorithmic techniques, particularly Machine Learning, to provide an insight from the data or to make informed decisions upon them.

- *Actuators:* Once a decision is made, it is communicated to an actuator (*e.g.*, drilling head and pressure valve) to enact the decision (*e.g.*, increase or decrease the pressure).

- *Security:* O&G is one of the most competitive industries worldwide and it has been the source of conflicts between many companies and even countries. That is why both the physical and cyber security of O&G is crucial and is considered as prominent elements of any infrastructure solution for smart O&G.

1.3.2 Edge-to-Cloud continuum and O&G industry

1.3.2.1 Cloud computing

Cloud computing is a concept that enables resources (*e.g.*, computing, storage, services) to be available as a service, on-demand, configurable, and also shareable [5]. Modern cloud systems provide diverse services in different levels, such as infrastructure as a service (IaaS), platform as a service (PaaS), software as a service (SaaS), and function as a service (FaaS).

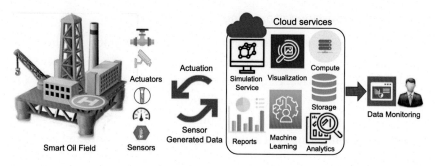

Figure 1.2 A wide variety of cloud services can be employed to store, process, and analyze sensor-generated data and to control industrial equipment in a smart oil and gas industry.

As depicted in Fig. 1.2, the smart O&G industry increasingly relies on cloud-based services that are hosted on remote Internet servers (a.k.a. cloud data centers). These data centers are utilized to store and process their data. According to Fig. 1.2, various sensor-generated data are sent to cloud providers to avail of different kinds of cloud services. Among these services, some of them send insightful decisions to actuators to close the automation control loop in the smart oil field. Cloud technology enables O&G companies to utilize various data-related and computational services (*e.g.*, machine learning and visualization) without the need to maintain any computing infrastructure. However, data privacy and security have remained a concern for such companies to fully embrace the cloud services. These security concerns have caused a small pause and hesitation in the adoption of cloud services, particularly, by major players in this industry. An alternative and more trustworthy approach is to store the data on an on-premise computing facility that is known as a *private cloud* (more recently called *fog computing*).

On the positive side, cloud systems' performance and ease-of-use are tempting for the O&G industry. For instance, one of the main users of data-driven cloud services is the North American shale industry that drills thousands of wells every year [6]. The scalability feature of cloud services helped the growing amount of data from these wells to be utilized efficiently, allowing the industry to expand remarkably. As such, various modern cloud-based data analytics services have emerged to help O&G companies to improve their operational workflows and make profitable decisions.

1.3.2.2 Edge-fog-cloud computing

Edge computing is defined as pushing the frontiers of computing applications, data, and services far from centralized nodes to the edges of a network. Unlike clouds that have a centralized nature and suffer from latency and security aspects, edge computing allows analytics and knowledge gathering to occur at the supply of the information. Accordingly, the core concept of fog and edge computing is bringing the computation close to the end-user (on-premises) or data generator. Hence, distance (and therefore latency) is a distinguishing factor to separate Fog/Edge computing from the cloud. Both fog (a.k.a. Cloudlet) computing and edge computing provide the same functionalities in pushing both data and intelligence to platforms situated either on or close to where the data originated from, whether that is screens, speakers, motors, pumps, or sensors. The aim of these technologies is not necessarily reducing the reliance on cloud platforms to analyze data. Instead, they can make more secure and faster data-driven decisions for latency-sensitive operations, and the rest of workload is still processed on the cloud.

The main distinction between edge and fog computing comes down to where the processing of that data occurs. Fog and edge computing appear similar since they bring the intelligence and processing closer to the data population. However, the critical difference between the two lies in where the computing power is placed. A fog environment places intelligence at the local area network (LAN) level. Alternatively, edge computing places intelligence and processing power in devices such as embedded automation controllers.

In sum, we can say the modern computing systems form a hierarchical structure (continuum), starting from at the device (*e.g.*, a sensor) level to edge (*e.g.*, a handheld processing machine), fog, and then cloud levels that complement each other. Fig. 1.3 demonstrates the Edge-to-Cloud continuum as a triangle where edge devices reside close to ending devices (bottom of the triangle) and cloud data centers are the furthest computing entity from end devices. Therefore, cloud computing incurs a high latency for edge and fog computing. Alternatively, cloud computing has a high availability in terms of elasticity and computing power, whereas, edge and fog devices are often located in the user's premises, hence, are more trustworthy than the cloud.

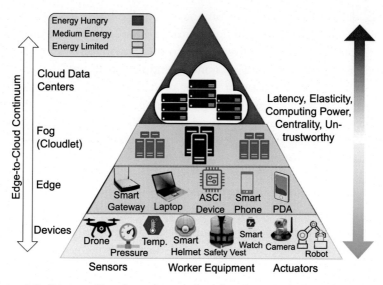

Figure 1.3 Edge-to-Cloud continuum for oil and gas industry. Four-tier computing with energy-limited devices at the bottom and energy-hungry devices at the top.

1.4 Smart O&G: data and software aspects

1.4.1 Big data in the O&G industry

The need for large-scale computing resources and cloud is because oil and gas industry produces a vast amount of data on a day-to-day basis. Following are three major sources (among many others) of generating such massive data in the O&G industry:

(a) Hydrocarbon reservoirs are typically located 5,000 to 35,000 feet beneath the Earth's surface. The only alternatives for locating and describing reservoirs are high-resolution imagery and costly well logs (after the wells are dug).

(b) Fluids must travel through complicated rock to reach the wellbore, and the fluids themselves are complex with many varied physical qualities. Learning about the unique characteristics of each oil well and analyzing the extracted fluid to efficiently process it requires massive data to be collected via sensors planted in the oil well and on the drill-head.

(c) Oil extraction poses environmental and human safety risks and avoiding it implies extensive sensor deployment in a vast geographical area to frequently collect data, hence, be able to quickly react to any environmentally pollutant emission.

Big data analytics helps streamline essential oil and gas processes in the upstream, midstream, and downstream sectors, including exploration, drilling, production, and delivery. Due to the growing use of big data analytics for the identification of nonconventional shale gas, the upstream sector is the most dominating source of data among all other sectors. Moreover, the oil and gas industry is becoming more unpredictable due to variable oil prices. As a result, in addition to the engineering team, business teams are largely embracing a data–driven strategy to predict the market and reduce their risks.

1.4.1.1 Big data in the O&G upstream

Numerous sensors in underground wells and surface facilities are used by oil and gas corporations to continuously collect data and monitor assets and environmental conditions in real-time. Sensors, geographical and GPS coordinates, meteorological services, seismic data, and other measuring equipment all contribute to the data volume. Specific applications are used to manage surveying, processing, and imaging, exploration planning, reservoir modeling, production, and other upstream operations utilizing "structured" data (e.g., databases). However, because much of this data is "unstructured" (or "semi-structured"), such as emails, reports, spreadsheets, images, voice recordings, and data market feeds, persisting them it in traditional data warehouses or querying and analyzing them on a regular basis is time- and cost-prohibitive. In this instance, Big Data-specific technologies are needed to handle the unstructured data.

1.4.1.2 Big data in the O&G midstream

The midstream sector of the O&G industry refers to everything needed to transport and store crude oil and natural gas before they are refined and processed into fuels and critical components for a broad range of everyday products. Pipelines and all the infrastructure required to transport these resources over long distances, such as pumping stations, tank trucks, rail tank cars, and transcontinental tankers, are considered parts of the midstream sector too. As such, midstream is known to produce big data from variety of sources and different big data analytics tools have been developed to process them. One instance is sensor analytics tools that are used by businesses to assure the safe logistics of their energy products. Another popular example is predictive maintenance tools that examine data from pipelines and tankers to detect anomalies (fatigue fractures, stress corrosion, seismic ground movements, and so on), allowing accidents to be avoided.

1.4.1.3 Big data in the O&G downstream

Given the increasing usage of product analytics solutions that aid refineries in standardizing the chemical composition of final goods, downstream is likely to be the second biggest sector (after upstream) to produce big data. Oil and gas industry may use big data predictive analytics and data management systems to help downstream companies simplify operations, enhance efficiency, decrease risk, and improve asset management by reducing downtime and maintenance costs for refining equipment. For instance, big data analytics has been used to undertake management optimization for refineries in Spain [7]. Moreover, downstream energy businesses can comply with environmental and safety laws via monitoring pipeline and refinery equipment in real-time. Cost-effective solutions can aid in the storage and management of the vast volumes of data generated by these applications. Big data also helps to estimate the demand for oil goods in the retail sales network and analyzing the price variations across different areas.

1.4.2 AI-based software systems in O&G

The smart O&G industry is part of the Industry 4.0 revolution that is mainly driven by artificial intelligence (AI), IoT, and state-of-the-art computing systems (*e.g.*, edge, fog, and cloud computing). Due to the vast implementation of smart sensors and actuators, a wide range and amount of valuable data are collected from various sectors of the O&G industry. These data can be analyzed via Machine Learning models to extract meaningful insights and knowledge that is beneficial to the industry and the environment. Therefore, in a broad sense, AI is the critical tool that with the help Edge-to-Cloud computing can transform sensor-generated data into an innovative and meaningful information and knowledge.

Data-driven methods offer a collection of tools and strategies to integrate various forms of data, assess uncertainties, identify underlying patterns, and retrieve relevant information. Data-driven software applications operating based on Machine Learning (ML) and Deep Neural Network (DNN) models [8–11] have been emerging as the fundamental pillars of the Industry 4.0 revolution [12–15], particularly, in remote areas where there is a need for real-time closed-loop automated processes, such as oil production control and emergency management systems [16–20]. The ML-based solutions are often in the form of microservice workflows with one (or more) critical path(s) that defines the overall application latency [16]. Handling these applications necessitates: (A) massive real-time data collection;

(B) seamless sensing data communications despite wireless link uncertainties; (C) reliable latency-constraint ML application execution in the face of unforeseen load surges (*e.g.*, during emergency situations); and (D) transparent (a.k.a. serverless) application deployment and resource provisioning.

Addressing these demands is challenging, especially in remote regions (such as offshore oil fields) with weak network connections and uncertain access to back-end cloud services. These limitations in communication and computing are especially important when a remote system must quickly adjust to an emergency scenario (*e.g.*, an oil spill) by processing enormous amounts of data in real-time to manage different aspects of the catastrophe. Although micro datacenters, a.k.a. *fog systems*, are used to address the computing needs of such distant systems, their capacities are often insufficient to handle the real-time data transfer and processing of the load spike [4]. We get back to this resource constraint challenge again in Section 1.4.4.

1.4.3 Digital twin: another data-driven applications in O&G

A digital twin is defined as a virtual replica of a real system. The sensors capturing data from their real-world counterpart can be configured to provide input to the twin. This enables the twin to stimulate the physical item in real-time and provides the management with the expected performance or other consequences [21]. DT is a data-driven application that takes advantage of the intelligence of Machine Learning, and the computing power of the cloud computing to achieve the dream of integrating all data to make the reality [22]. Unarguably, data is crucial for analytics, prediction, and automation in a DT system. The data has to be of high quality, validated, and referenced to create a usable twin. To function in real-time, the DT requires existing data and models to remain updated and valid [23]. The DT systems can bring about operational excellence via empowering the operators and managers in the O&G industry to convert massive volumes of data into insights that can make asset failure foreseen, and hidden income possibilities can be disclosed.

O&G industry can potentially utilize digital twin technology to visualize a process or system, generate an analytical what-if model, anticipate the future, or aid with idea selection via providing complete visuals. It may also shorten the time necessary to analyze data and produce prospects by employing AI-assisted picture recognition. These advantages have motivated several oil and gas players to invest in the DT technology [24]. APEX, a complex simulation and surveillance system created by BP, has been used to generate virtual models of its production systems. APEX enables BP to test and prepare modifications and interventions in the DT before implementing them in the real world. As a monitoring tool, it detects problems before they affect the productivity. Unlike the traditional simulations, APEX can run the simulations in a near real-time manner, and the impact of potentially dangerous decisions can be pre-evaluated in a virtual world. For instance, Petrobras implemented the DT technology in 11 of its refineries, enabling them to assess and adjust operating parameters in near real-time. This enabled the organization to improve operational efficiency and process optimization while saving around $154 million. As the digital twin technology progresses, its involvement in the O&G industry is projected to become more critical.

1.4.4 Edge-to-Cloud for AI and other data-driven applications in smart O&G

Typically, an Edge-to-Cloud continuum with local appliances connected to sensors is difficult to operationalize, due to the diversity of sensors that communicate via heterogeneous protocols, such as Modbus, CAN bus, PROFINET, and MQTT [25]. The scale of deployment, including hundreds of departments and thousands of oil rigs, usually makes it even more challenging. Moreover, the next-generation of cloud-native applications require new specifications, configuration, and various machine learning (ML) frameworks. Applications need to be compatible to run on a range of devices with various computing resources (*e.g.*, CPU, various types of GPU, ASICs, and FPGAs). In addition, the human element of IT operational technologies, developers, and data scientists all need to come together to operate the IoT application deployed in the Edge-to-Cloud continuum. As such, we can summarize the main challenges of the Edge-to-Cloud continuum as follows:

1. Establishing communication among IoT devices, edge, and the cloud in the large-scale.
2. Communication costs of wireless communication solutions.
3. Getting access to computing power with guaranteed performance when needed.
4. Challenges with slow, interrupted or nonworking wireless connections.
5. The need for real-time operation of ML-based and other data-driven applications (*e.g.*, digital twin).
6. Data integrity and privacy across Edge-to-Cloud systems.

As we can see, the challenges of Edge-to-Cloud continuum for the O&G industry are diversified, complex, and different from the legacy solutions. Hence, both petroleum experts and technical specialists are the core driving forces to devise profitable environment-friendly solutions for the smart O&G industry. Nonetheless, it is rare to find scholarly articles, research papers, and books that address the intersection of petroleum and computer science areas, thereby, reducing the gap for understanding the problem space and the proposing solutions.

To overcome these challenges and unfold the complexity of these systems for data-driven applications, we can envision a smart O&G system as a cross-layer framework that includes (wireless) networking at the lowest level; and edge-fog-cloud computing at the platform level, and multiple ML applications to support emerging industry 4.0 data-driven applications in the oil fields at the application layer. Below, we describe two concrete use cases that are representative of the scenarios we aim at considering in this book in which ML-based O&G applications are provisioned using edge-cloud systems.

Sample use case 1: Federated fog in remote O&G fields. Fig. 1.4 shows one scenario where multiple oil fields are located in a remote (offshore) area with low network connectivity. The cross-layer framework that can handle different data-driven applications of this system operates based on federating the nearby computing systems fog systems (as opposed to connecting to remote cloud services) over high-capacity wireless links. In this case, the physical and platform layers provide the user applications with the illusion of accessing infinite cloud-like resources and the ability to seamlessly handle the surge loads of multiple data-driven applications that can occur in scenarios like oil spill detection, seismic data analytics, etc.

Sample use case 2: Pipeline inspection. Let us consider a scenario where 4K drone-mounted cameras can collect hundreds of gigabytes of data per hour. The current method of analyzing data is to transfer the massive data to the cloud datacenter, which is cost-prohibitive and impractical, especially, if the analysis is of real-time nature. This scenario is presented in Fig. 1.5 where the drone-based inspection system could use multi-stage value extraction. The O&G pipelines can be thousands of miles long and pass through an immense landscape. Pipe sections are generally fitted with analog gauges and sensors to measure pressure, temperature, flow, and other critical metrics. By employing an edge AI-enabled surveillance drone, shown in Fig. 1.5, to capture these analog gauge images, it is possible to separate the gauge images and transfer only that critical information to the next computing tier. Here, data preprocessing (*e.g.*, image separation) can be done on the drone itself. Then, in Step 2, the data undergo

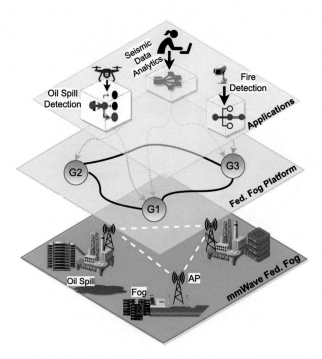

Figure 1.4 A layered view of an example smart oil field in a remote area with low or no access to central cloud infrastructure. In the highest layer (application later), various data-driven machine learning applications can be executed in this smart oil field. In the second (middleware) layer, the computing infrastructure needed to execute this resource-hungry applications can be provisioned via a federated nearby fog computing systems located in nearby oil rigs. At the lowest (physical) layer state-of-the-art millimeter-wave wireless communication can be utilized to enable high-bandwidth communication across the federated systems.

a lightweight processing (*e.g.*, via quantized ML [26]) at the edge tier to identify images that are suspect of showing some form of leakage. In Step 3, only the identified images are transmitted to the back-end cloud for in–depth analysis and accurately identifying leakage.

1.5 Roadmap of the book

This book explores the O&G industry through its three main sectors from supply chain management perspective i.e. upstream, midstream, and downstream. In Chapter 1 (current chapter), we discussed the overall scope,

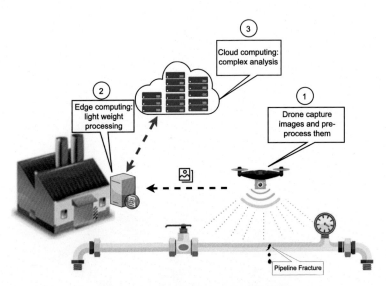

Figure 1.5 Drone-based inspection scenario where the drone captures images and real-time analysis is performed on the edge computing resources, whereas, long-term analysis is performed on distant cloud computing facility.

objectives, and challenges of the O&G industry. We briefly discussed the advanced computing technologies, and their utilization examples. A brief summary of the other chapters is provided too.

Chapter 2: Upstream

This chapter explores the upstream sector of oil and gas that mainly consist of exploration, and production. Along with a conventional taxonomy of different aspects of upstream, we provide a second taxonomy for computing in the upstream sector that reflects the scope of computing technology adopted to oil and gas. After that we traverse both taxonomies and especially explain in details on the computing taxonomy, as it is the main objective of this book. This chapter clearly explains the upstream operations from exploration surveys to production wells and the scopes of using state-of-the-art computing technologies to improve the oil and gas production, while maintaining the environment safe at the same time.

Chapter 3: Midstream

This chapter explores the midstream operations of O&G that mainly includes transportation, storage, safety, and security. It begins with demonstrating taxonomy of the midstream sector from the petroleum aspects, and then bringing the computational perspective into the picture. The chapter investigates various solutions of midstream that can be made efficient via utilizing smart computing solutions offered by IoT, Edge, Fog, and Cloud computing platforms.

Chapter 4: Downstream

The downstream sector of O&G is elaborated in this chapter. Our main focus in this chapter is to address challenges of downstream sectors, and to provide smart, and efficient solutions utilizing Cloud technologies. For that purpose, we introduce the main operation of a refinery in processing crude oil. This chapter is followed by a taxonomy of downstream that is divided into three subcategories, namely refining, distribution, and safety. The scope of utilizing state-of-the-art computing technologies is addressed through several application cases. In addition, the chapter provides different use cases that utilize IoT devices, sensors, and controller.

Chapter 5: Threats and side-effects of smart oil and gas

The smart technologies are changing modus operandi operations in various sectors of O&G. However, smart technologies in such a large-scale deployment with numerous connected devices depending on one another are prone to ambiguity and uncertain situation. In such a system, if a step produces any undesired output, it can be propagated in the whole system and lead to unintended consequences. In addition, appropriate human-machine interaction is yet to be flawlessly established in the current smart systems and this can be another threat for modern industrial systems. Considering these important challenges, we dedicate Chapter 5 of this book to investigate these threats and discuss possible solutions for them.

Chapter 6: Designing a disaster management system for smart oil fields

Smart oil fields utilize a variety of sensors (*e.g.*, pipeline pressure, gas leakage, and temperature sensors) to solve the challenges of oil extraction that generates terabytes of data every day. The data is sent to cloud datacenters through high-latency and unreliable satellite connections. Although edge computing may be used on the oil rigs to execute latency-sensitive operations, its processing capacity is limited during disasters when several coordinated actions must be carried out within a short period of time.

Hence, a robust smart oil field that operates on the federation of edge computing systems provided from nearby/mobile mini datacenters can be established. The smart solution accomplishes resilience by capturing uncertainties that exist in the federated environment's communication and computation while distributing urgent tasks in such a way that their chance of being completed on time is maximized. Compared to traditional oil fields, the evaluation findings show that this solution has a considerable performance gain.

Chapter 7: Case study I: Analysis of oil spill detection using deep neural networks

Environmental pollution is one of the world most pressing issues, and the oil spill in several historic instances has proven to be a major source of environmental pollution. Early discovery of an oil spill can be decisive in preserving the environment, and recovery technologies are especially effective to save the marine life. Satellite pictures are the primary source of these detection systems and are captured using Synthetic Aperture Radars (SAR). Oil spills are captured as black dots in SAR sensors that create difficulty in distinguishing between the actual oil spill and a look-alike. To identify and characterize SAR black dots, researchers used a wide variety of methodologies. However, because most of them use a customized dataset, the results are not comparable. As a result, deep convolutional neural networks (DCNNs) are proposed as a reliable approach for this problem. They can use SAR images to detect oil spill along with other critical classes efficiently. Therefore, a study based on DCNN has been conducted to identify oil spills and related classes (*i.e.*, ship, land, and look-alike) on a real SAR dataset. The main focus of this chapter is to show a case study on the potential impact of the DNN model configurations on various computing platforms (*e.g.*, fog, edge, and IoT) in detecting the oil spill.

Chapter 8: Case study II: Evaluating DNN applications in smart O&G industry

Industry 4.0 relies on cloud-based Deep Neural Network (DNN) applications that perform latency-sensitive inference operations. However, the inference time of DNN-based applications is stochastic due to multi-tenancy and resource heterogeneity, both of which are intrinsic to cloud computing settings. If not recorded, such stochasticity can result in poor Quality of Service (QoS) or even disaster in essential industries like oil and gas. To make Industry 4.0 resilient, solution architects and researchers must understand the behavior of DNN-based applications and capture the inference uncertainty. In the second use case, we demonstrate a descriptive analysis of inference time from two viewpoints to enable a safe and secure smart solution for the O&G industry. First, this chapter statistically models the execution time of the four categorically different DNN applications on both Amazon and Chameleon clouds. Then, a resource-centric analysis is provided for these applications on heterogeneous machines in the cloud. The findings of this case study are helpful in developing robust solutions against the stochastic inference time of DNN applications in the cloud, offering a higher QoS to their users and avoiding unintended outcomes.

CHAPTER 2

Smart upstream sector
Smartness in upstream sector of the oil and gas industry

Chapter points

- The scope of smart technologies in various fields of upstream sector is explored.
- The taxonomy of upstream sector from petroleum, and computing perspective are demonstrated.
- The complex upstream oil and gas operations are investigated from the computing perspective, and feasible solutions are proposed with pictorial explanation.
- The utilization of various smart technologies including IoT, Fog, Edge, and software solutions from machine learning, and artificial intelligence are investigated.

2.1 Introduction and overview

Today's O&G industry involves a lot of computation in its day-to-day activities. As mentioned earlier, having various cyber-physical devices (*i.e.*, sensors, actuators, smart gateway), the O&G industry produces an immense volume of data that needs to be processed, stored, and analyzed for capturing valuable business and operational insights. Various machine learning and deep learning models can be trained with this data and later be utilized in automation systems, and other operations of the upstream sector. Technologies such as IoT, robotics, cloud, edge computing, and artificial intelligence (AI) are jointly transforming the upstream sector of oil and gas industry to a digital model.

Upstream activities can be carried out more efficiently when functional teams involved in finding and producing crude oil and natural gas work together from a standard playbook. Two digital solutions can help oil and gas companies to streamline and integrate their upstream operations, namely Digital Twin [27], and Single Engineering Platform [28]. A digital twin is a booming concept in Industry 4.0 and is defined as the digital representation of physical components. The digital twin is a virtual representation of a machine, process, or other real things. It is based on real-world data to

IoT for Smart Operations in the Oil and Gas Industry
https://doi.org/10.1016/B978-0-32-391151-1.00011-3

precisely describe the behavior of a procedure or item. With advancements in artificial intelligence and machine learning, digital twins have become more complex and accurate. For instance, drilling rigs for oil and gas are extremely complicated systems with numerous operations that must work 24 hours a day in difficult weather conditions. Hence, a digital twin (*i.e.*, 3D view) of the hydrocarbon reservoir can enable the prediction and advanced planning capabilities for the well construction processes. In the construction phase, the asset's digital twin makes it possible for the supervisor to not only coordinate work orders more safely and efficiently, but also maximize technicians' productivity with an overlay of augmented reality features, such as hidden mechanical and electrical components.

Leading O&G companies (*e.g.*, BP, Chevron) are utilizing digital twin [29] for asset optimization, extraction operation monitoring, and improving reliability. According to the BP experience, the digital twin has aided engineers in optimizing assets and identifying possible hazards before they occur. In addition, the digital twin enables optimization process to be completed in less than 20 minutes, which has been otherwise taken more than a day. As a result, it reduces the time for insight while also freeing up critical engineering time. From historic oil fields to modern natural gas developments, Chevron has been incorporating various digital technologies into its operations [30]. According to the corporation, billions of dollars are being invested in "strengthening its basic business, keeping its new assets online, and boosting output." One of the keys to assure optimal production is to improve dependability. "It entails expanding and automating our processes by leveraging current, developing, and yet-to-be-developed technology and workflow advances," according to Chevron. In addition, it entails creating better data that ML software can translate into valuable information in real-time, allowing it to operate more securely, reliably, and efficiently while also lowering costs, recovering more resources, and better managing risk. A Single Engineering Platform reduces engineering costs by enhancing collaboration and reducing feedback cycle times. Engineering and software companies cooperate to build integrated, 3D-centric and cloud-based platforms to enable ecosystem partners to work smarter and faster.

The upstream sector is responsible for all activities that lead to the discovery and production of crude oil or natural gas. The initial step in the upstream sector is to explore a wide area for sufficiently large and feasible crude oil or gas resources. If the output of the first step is satisfactory, the next step is to plan and drill the engineered wells to reach the hydrocarbon reservoir for stable oil and gas production. Finally, successfully well drilling will result in the oil production phase. While the upstream sector is widely known as the Exploration and Production (E&P) sector, it is better to have a more granular categorization for further investigations. Fig. 2.1 shows more granular activities in the upstream sector. The essential activities in each category (*i.e.*, Exploration and Production) are demonstrated in this taxonomy. In the next three subsections, these three categories are elaborated.

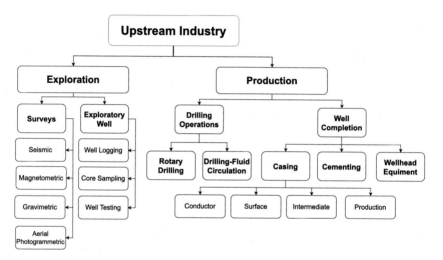

Figure 2.1 Taxonomy of the upstream sector of O&G industry. The first phase of O&G industry is Exploration. This subsection shows the most important activities that finally leads to discovering a sufficient and economical oil and gas reservoir. The next phase is Drilling the well. Drilling activities and operations are responsible for designing and constructing wells appropriate for oil production. The last phase in upstream sector is related to oil production. In this phase, wells that are drilled in previous phase are becoming ready for oil production. Monitoring and maintenance activities are important activities in this phase.

Considering all the smart operations of the upstream sector, a well-defined taxonomy based on the computing perspective has been developed to identify different smart activities in the upstream sector. This taxonomy is provided in Fig. 2.2.

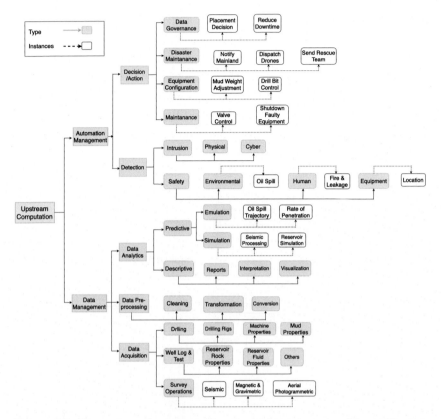

Figure 2.2 Taxonomy of computation in upstream sector. The main two branches are data management and automation management. The solid line indicates the concept or type, and the dash line indicates the instance (*i.e.,* example) of that particular type.

2.2 O&G exploration

Exploration is defined as the efforts made to find where oil and gas reservoirs exist, but it is also economical for production. Various surveys are conducted by geologists, geophysicists, and petroleum engineers to indicate the existence of a hydrocarbon reservoir and determine its characteristics. It might also be required to drill an exploratory well to access the reservoir for a more accurate analysis of the rocks and fluids formation. These surveys are briefly explained in the following subsections.

According to the strategic planning timeline, the exploration subsector can be of various lengths. For example, a typical timeline for exploration according to different operations is shown in Fig. 2.3.

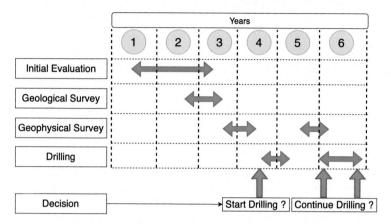

Figure 2.3 Timeline of exploration stages of upstream sector up to drilling. First 3 to 4 years of exploration stages includes primary evaluation and various survey operations. Based on initial surveys exploratory well drilling starts and collect logs while drilling. Depending on the collected logging data, well drilling and well completions start and continue till production phase.

2.2.1 Survey studies for potential reservoir

In exploration, surveys are essential operations based on planned drilling and production phases. Surveys are conducted to evaluate the subsurface formations and fluids. There are many surveys the O&G industry performs to identify or locate trapped hydrocarbon beneath the earth's surface. The major surveys are as follows:

- Magnetometry survey
- Gravimetric survey
- Seismic survey
- Aerial photogrammetric survey

Typically, these surveys generate data for modeling/understanding subsurface formations by geophysicists. Surveys are generally batch task types from the computational perspective. Surveys in exploration play a vital role in deciding where to start the extraction operation. Therefore, precision and time period for performing any survey are important factors influenced greatly by the computing activities. The remainder of this section will explore the survey operations in detail.

2.2.1.1 Magnetometry survey

The magnetic survey is one of the most popular oil and gas exploration methods. A sketch diagram for geophysical interpretation of gravity and magnetic anomalies is presented in Fig. 2.4. The concentration of ferromagnetic minerals in geological formations changes the magnetic properties of underlying rocks and minerals. Thus, measuring anomalies in the geomagnetic field can reveal the physical properties of the subsurface formations. An aerial survey may be used to explore a vast area while conducting a preliminary evaluation of a region's geological features and characteristics. A magnetic sensor is fixed to the tail of an aircraft to measure the magnetic field anomalies. The airborne magnetic survey is a rapid and comparatively low-cost survey. A ground survey is appropriate for exploring relatively small areas that an airborne survey has already determined.

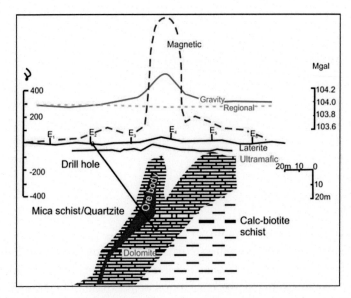

Figure 2.4 Geophysical interpretation of gravity and magnetic anomalies, and confirmed by drill testing of rich sulfide orebody in Rajasthan [31].

2.2.1.2 Gravimetric survey

The gravimetric survey is a geophysical method that identifies anomalies (as depicted in Fig. 2.4) in the strength of the gravitational field to describe the subsurface rock formations. An underlying rock formation with a comparatively lower density causes negative anomalies in the gravitational field, thus, reducing the downward gravitational force. Measurements of gravitational variations can be carried out on land and sea. While traditional terrestrial gravimetry provides high-resolution data, it is restricted by severe environmental conditions like high mountains or jungles. In this situation, satellite-based gravimetry with limited spatial resolution can be utilized to determine the gravitational variations. Moreover, Airborne and shipborne gravimetry can be employed to fill data gaps of the traditional gravimetry on the ground and the satellite-based methods.

2.2.1.3 Seismic survey

The seismic survey is another form of geophysical survey that is used to describe the subsurface structure (*e.g.*, layers and discontinuities). The seismic survey is based on the theory of elasticity. In this survey, the response to the elastic waves generated artificially by explosions or vibrations determines the elastic properties of the subsurface materials as depicted in Fig. 2.5. To be more specific, the generated seismic waves reflected back from the interface of different layers of rocks at various depths in the earth. Then, a certain type of sensors on the ocean (subfigure (a) of Fig. 2.5) surface or land (subfigure (b) of Fig. 2.5) surface (hydrophones for marine acquisition and geophones for land survey) receive and record the arrival time and amplitude of the reflected waves. Finally, the computer-assisted analysis of the collected data reveals the sedimentary structures of the subsurface and shows the location of porous layers within these structures.

2.2.1.4 Aerial photogrammetric survey

The technique used to obtain measurements, maps, or 3D views of the area of interest from a set of photographic images is called photogrammetry. Aerial photogrammetry uses special cameras mounted in airplanes to take photos. The Photographs, along with photogrammetry techniques, finally provide a three-dimensional view of the earth.

All these surveys help the geologists and petroleum engineers to predict the presence and locate the oil reservoir. Next, the O&G industry starts drilling exploratory wells for further analysis. According to our taxonomy

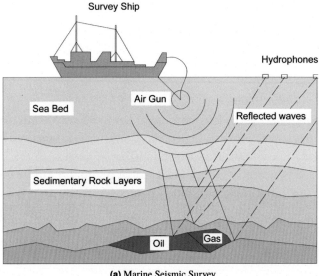

(a) Marine Seismic Survey

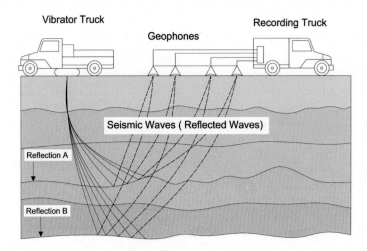

(b) Terrestrial Seismic Survey

Figure 2.5 The figures depict the seismic survey on land and sea. The artificially generated waves are reflected back from different layers in subsurface. Then, geophones (on land) and hydrophones (on sea) receive and record the amplitude and arrival time of the reflected waves. The analysis of the recorded data by geophones or hydrophones can reveal the structure of the subsurface layers (*e.g.*, discontinuities in layers).

for upstream, the exploratory well construction following drilling consists of oil well logging, core sampling, and well testing. These three operations are described in the following paragraphs.

2.2.1.5 Well logging

A well log is a record of geological formations penetrated by the borehole with respect to both time and depth during the drilling process. Logging tools are inserted into the well to measure various properties of the subsurface formations. Different types of well logging are used for different purposes. For example, acoustic or sonic well logs identify the formation porosity. Alternatively, spontaneous potential (SP) well logs are useful to delineate formation type. Generally, as a consequence of well logs, thickness and formation types, porosity, permeability, water saturation, temperature, and reservoir pressures can be measured. It is noted that the main goal of well logs is to find whether a well is economically feasible or not. In the case of positive analysis, the O&G industry will move towards preparing the well for the production phase.

2.2.1.6 Core sampling

A core is a cylindrical length of rock that has been cut by a hollow bit and driven up into a tube (core barrel) that has been connected to the hollow bit. After being transported to the surface, the core barrel is opened, and the core is taken for geologic investigation. In subsequent steps, the sample is utilized to evaluate the physical parameters of the formation (*e.g.*, porosity and permeability), as well as the geologic age and productivity of the oil reservoir.

2.2.1.7 Well testing

In contrast to other geophysical surveys and well logging, a well test is used to directly measure flow rate and pressure. Furthermore, the correlation between pressure and flow rate is used to determine the flow conductivity of the reservoir [32]. Well testings can appropriately investigate the performance of the well. Oil and gas flow rates, gas to oil ratio, flowing downhole, and surface pressure and temperature are some of the parameters recorded in a well testing process [33].

2.2.2 Smart O&G exploration: how can computing help?

The introduction of the fourth Industrial Revolution (Industry 4.0) transforms many operations of the O&G industry. These transformations are directed toward the unified aim of improving and optimizing existing processes and technologies. Along with the industry 4.0 revolution, cyber-physical devices (*e.g.*, smart sensors, actuators, and switches) are installed in many sectors (*e.g.*, exploration and drilling) of the O&G industry as a part of digitization. For example, in well exploration, 4D seismic data captured from sensors help geologists to analyze the throughput from a potential reservoir and provide decision pointers. In the digitization process, the O&G industry adopts many digital transformations discussed briefly in the following subsections.

2.2.2.1 Machine learning in geoscience for oil and gas exploration:

In the oil and gas exploration, various surveys (*e.g.*, seismic, gravimetric, magnetometry) are conducted that utilize smart sensors for collecting data. The exploration data is then analyzed by the geoscientist to bring out meaningful insights. Machine learning algorithms are currently used widely in geoscience modeling to help geoscientists in inferring the estimation of hydrocarbon volume underneath the surface of earth [34]. Climate change is one of our generation's defining issues. The World Health Organization (WHO, 2014) recently published a report concluding that air pollution caused seven million deaths worldwide in 2012, with atmospheric aerosols playing a substantial role. However, we still do not have an actual aerosol abundance product for the boundary layer due to the significant limitations of remotely detecting the atmospheric boundary layer. One method might be to develop a Virtual Sensor that provides an accurate boundary layer atmospheric aerosol product with an accompanying uncertainty by merging roughly 50 tremendous remote sensings and Earth System modeling products using multivariate nonparametric nonlinear ML technique.

2.2.2.2 Using cloud data centers to archive exploration data

The wave of digitization forces the O&G industry to preserve the exploration data in a secure place for further analysis. The preservation of subsurface data is challenging due to its volume and old unstructured data (i.e., images of the log file, video recording) on predating computer storage [35]. For example, a modern seismic survey generates thousands of terabytes of data, whereas state and federal repositories all together hold hundreds of miles of core [36]. In addition, millions of digital and paper records are stored at state geological surveys. For instance, the Kansas Geological Society library maintains over 2.5 million digitized well logs and associated records for the state [37]. Hence, cloud data centers are providing storage solutions for the O&G industry. Moreover, cloud data centers provide structured database services to store various data formats.

2.2.2.3 Automation in exploration data acquisition

One of the digitization aspects of the O&G industry is the implementation of automation in various sectors. For example, in oil and gas exploration, automatic data acquisition and transmission of the data to the processing unit has improved the efficiency of exploration operation. Furthermore, with the help of real-time data acquisition and on the spot processing utilizing edge computing technology has improved the production rate and saved a significant time for starting the production [38].

2.2.2.4 Virtual reality in seismic imaging

According to Deloitte's Digital Operations Transformation (DOT) framework, [39], seismic imaging—a technology that has been used in the industry for over 80 years to evaluate and image new and complicated subsurface formations—is essentially at an advanced degree of data analysis and visualization. The analytical lead of this segment is explained by the standardization of geological data and formats, company investment in innovative algorithms, and the movement toward high-performance computers that can diagnose a significant amount of geoscience data in a few seconds. For example, ExxonMobil utilizes seismic imaging to anticipate the distribution of cracks in tight reservoirs, allowing it to improve flow and well location.

Virtual reality is being used in 3D modeling to make the visual experience more realistic, which may aid geoscientists in inferring numerous hidden truths. One aspect to consider is incorporating virtual reality elements into seismic imaging to improve the spatial perception of 3D objects. For instance, researchers at the University of Calgary use virtual reality, augmented reality, and advanced visualization methods to aid Canadian petroleum producers in utilizing stream-assisted gravity drainage (SAGD) better maintain their complex reservoirs by introducing simulations in a realistic 3D world.

2.3 Well production

2.3.1 Drilling operations

The main key elements to have a successful drilling in oil or gas production are as follows:

- weight on the bit (WOB) or the force applied downward on a drill bit
- drill bit rotation
- drilling-fluid circulation

The weight on the bit is continuously monitored to maintain its optimal value. The excessive force can cause the buckling problem, while the less than optimal value reduces the rate of penetration (ROP), thereby, decreasing the performance of the drilling operation. As such, continuously adjusting the WOB to its optimal values is crucial during the drilling operations. In addition to WOB that provides the downward force, the drill bit has to be rotated at a certain speed to cut the rock and move forward through the rock.

The heat and cuttings generated by the drilling operations (*e.g.*, drill bit rotation) should be continuously removed during the drilling operation. To this end, drilling fluid flows through the whole path of the drilling hole to dissipate the heat and remove the cuttings. In addition, the drilling fluid balances or overcomes formation pressures in the wellbore. Moreover, circulating drilling fluid forms a thin, low-permeability filter cake on the wellbore walls. The filter cake minimizes the liquid loss into formations and helps stabilize the wellbore until the casing is set with cement. Therefore, the success of the drilling program mainly depends on the successful operation of the drilling fluid. Furthermore, the proper selection and maintenance of the drilling fluid would reduce the overall cost of the drilling program.

2.3.2 Well completion

Well completion is the process of making a well ready for production. This process mainly involves the following steps:

- Preparing the bottom of the hole to the required specifications.
- Running in the production tubing and its associated downhole tools.
- Perforating and stimulating as required.

Casing is the steel tubes that are set inside the drilled well to serve the following functionalities:

- Stabilize the borehole
- Isolate the wellbore fluids from the subsurface formation

- Control the formation pressure with the help of blowout preventers (BOPs)

The casing is classified into many sorts based on its function and diameter. Various kinds of casing are used at different depths of the well, including conductor, surface, intermediate, and production casing. Cementing is the following stage once the drilling operations are completed. Cement is used to fill the annular area between the casing string and the drilled hole. The primary objectives of cementing tasks are to increase formation stability and zone isolation.

It is worthy to mention that a well cannot be drilled to its target depth without intermediate steps of running, casing, and cementing. The cyclic process of drilling, casing, and cementing is illustrated in Fig. 2.6. Casing and cementing processes improve the strength and stability the well. In addition, these processes help to minimize damages to the formation by isolating the drilling fluid from the formation.

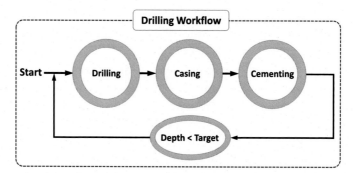

Figure 2.6 Drilling a well to a certain depth is an iterative process, each iteration includes drilling, casing and cementing. When the drilling rig drills the well to a certain depth, the casing string is run through the well and cemented. Casing prevents collapse of the wellbore and isolates the formation fluid from the borehole fluid.

Eventually, to make a connection between the wellbore and the formation, perforation guns are run into the wellbore to blast holes in the casing.

2.3.3 Smart solutions in well production

2.3.3.1 Machine learning to predict drilling rate of penetration (ROP)

The performance of drilling significantly depends on the continuous monitoring of drilling subtasks (*e.g.*, weight adjustment of drill bit, and drill bit rate of penetration). Anomaly in the drilling variables can lead to various problems, such as bit balling and broken drill bit, that can be predicted beforehand via utilizing digital log data of the drilling. With the help machine learning models utilizing the digital log data of drilling the ROP can be predicted and the weight adjustment of drill bit can be performed in real–time [40] and [41].

2.3.3.2 Predicting drilling fluid density

To reduce drilling fluid losses through the formation, it is necessary to understand the rheology of drilling fluid at wellbore conditions (high pressure and high temperature). The lack of a general model for predicting drilling fluid density under the conditions studied hampers the drilling fluid loss control efficacy. However, a reliable machine learning model for estimating drilling fluid density (g/cm^3) at wellbore conditions may be constructed with the digitalization of drilling data. In this case, a high–performance model for forecasting drilling fluid density was suggested using a combination of particle swarm optimization (PSO) and artificial neural network (ANN). Furthermore, two competitive machine learning models were used, including the fuzzy inference system (FIS) model and a combination of the genetic algorithm (GA) and the FIS (dubbed GA–FIS) approach. Finally, data samples from the available literature were utilized to build and test the prediction models. Compared to other intelligent approaches analyzed in the research work [42], the PSO-ANN model results into an acceptable performance. As such, the PSO-ANN model is considered an appropriate model to accurately predict drilling fluid density (g/cm^3, or PPG) under HPHT conditions.

2.3.3.3 Drilling fluid optimal pressure estimation

The circulation of drilling fluid dissipates heat and removes the cuttings generated during the drilling operation. The drilling fluid is then pumped through the drill string. The pressure of drilling fluid static or its circulation density equivalent must also be monitored. The optimal pressure is the value to neither damage the formation nor have cross–flow (kick).

2.3.3.4 Intelligent decisions to mitigate lost circulation

Among various challenges of drilling, lost circulation is one of the significant issues in drilling a well. This problem stretches the down–time and operational cost and causes several hazards (mainly, wellbore instability, and pipe sticking) related to safety. It is unfortunate that there is no consistent and specific solution for resolving this lost circulation. Hence, scholars and researchers focus on monitoring drilling operation parameters and fluid characteristics demanded to analyze the lost circulation problem properly. Abbas et al. in [43] proposed a model to predict the lost circulation solution for vertical and deviated wells utilizing a machine learning method, namely support vector machine (SVM). This research aims to create an expert system that can screen drilling operation parameters and drilling fluid properties to accurately identify the lost circulation problem and provide the best remedy. Field datasets were obtained in the first stage from 385 wells drilled in Southern Iraq from various fields. The relevance and order of the input factors that impact the lost circulation solution were then determined using the fscaret package in the R environment. Artificial neural networks (ANNs) and support vector machines were used to construct novel models to estimate the lost circulation solution for vertical and deviated wells (SVM). The use of machine learning algorithms might help the drilling engineer make an informed choice on properly treating lost circulation.

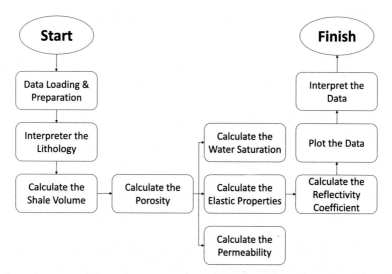

Figure 2.7 Data flow diagram of well log data interpretation for characterization of an oil or gas reservoir from data acquisition to eventually calculate reflectivity coefficient. This coefficient is used to identify a hydrocarbon bearing zone.

2.4 Smartness in the upstream sector

2.4.1 Overview of dataflow in upstream

In the upstream sector sensor generated data are acquired, preprocessed and finally with some analytic it performs automatic decision making with actuators. This whole process can be described as a workflow of data in the upstream sector that is depicted in Fig. 2.7. In this figure, an example of characterization of an oil and gas reservoir from reflectivity coefficient is demonstrated. The operation starts with data loading and preprocessing for interpretation of lithological characteristics of reservoir. Then, the shale volume is calculated and used to determine the porosity. After that, water saturation, elastic properties, and permeability are estimated using the porosity estimation. Finally, the reflectivity coefficient is calculated and utilized in plotting for interpretation of characteristics for analyzing oil and gas reservoir.

2.4.2 Smart sensor data acquisition

In accordance with the upstream sector taxonomy (Fig. 2.2), data acquisition branch from data management subsection has three subbranches that are defined in this work as: survey operations, well log, well tests, and drilling. In this section, we explore the computational aspects of these subbranches consecutively.

2.4.2.1 Data acquisition in survey operations

Exploration sector of upstream has many survey operations among which seismic, magnetic, and gravimetric are major operations. Hence, we explore these survey workflows and the way data are acquired in these operations:

1. **Seismic:** In this type of survey, sound waves are bounced off underground rock formations and the waves that reflect back to the surface are captured by recording sensors. Analyzing the time and the returned waves provide valuable information about rock types and possible gases or fluids in rock formations. The three following phases of the seismic surveys require computer processing:
   - Data Acquisition—Receivers capture geophone (onshore) or hydrophone (offshore) sound waves.
   - Processing—When these waves are printed, the sound patterns appear as "traces" of the subsurface.
   - Interpretation—Each individual wavelet's path reveals a specific contour of the geologic structure. The pattern of these wave gen-

erators and their recording devices can create many complexities of scientific data (*e.g.*, 2D, 3D seismic images). An example of seismic processing in cloud is represented in Fig. 2.8

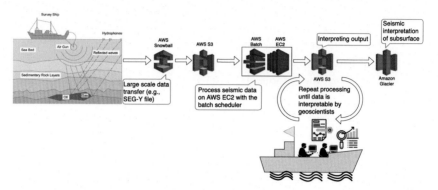

Figure 2.8 Seismic image processing in cloud from data acquisition to processing.

The main purpose of this survey is to find the location and size of oil and gas reservoirs. Collecting and processing the data can take 12–18 months. This operation is considered as batch jobs in the high performance computing environments. A pictorial representation for data acquisition, process, and interpretation of this type of survey is depicted in Fig. 2.9. According to this figure, seismic survey data are sent to cloud server for physical seismic modeling. Finally, a visual representation in form of 3D map is developed for further analysis.

2. **Magnetic survey:** In this type of survey, the magnetic effects produced by varying concentrations of ferromagnetic minerals, such as magnetite, in geological formations are measured. Magnetometer is

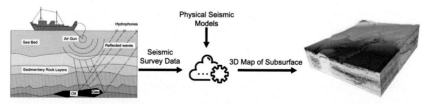

Figure 2.9 Acquired seismic data are used as the input to the physical seismic model that is processed in the Cloud. The processing output is a 3D seismic model of the subsurface, created based on the seismic images. The 3D output image is obtained from [44].

a specially-designed magnetic compass that detects minute differences in the magnetic properties of rock formations. Here, magnetometer (airborne magnetic unit) used to capture the data. One of the widely used magnetic instruments is the vertical magnetometer. The following three phases of the magentic surveys require computer processing:

- Data acquisition: The survey explores the local magnetic field characteristics with magnetometer.
- Data processing: Collected data are processed to develop magnetic image for further analysis.
- Interpretation: From the magnetic image, magnetic anomalies are identified to understand the storage of hydrocarbon beneath the earth surface.

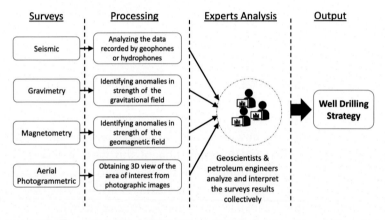

Figure 2.10 The collective information obtained from the surveys is interpreted and analyzed by geophysicists and petroleum engineers. Finally, they can determine the feasibility and the economic efficiency of the oil production from the explored area.

3. **Gravimetric survey:** The gravity field (g) of earth is affected by the density of different types of rocks. By mapping these differences, certain rock formations can be located. Here gravimeters are used to capture the changes in the earth gravitational field. The principle of a gravimeter is the use of a very sensitive spring (as noted in Fig. 2.11) and weight system attached to a beam. As gravity increases, the weight is forced downward that stretches the spring. Gravimeters measure the ambient gravitational field at any specific point or station. The three phases of this survey that require computer processing are as follows:

- Data acquisition: All gravity surveys measure the vertical component of gravity, that is gz. The gravimeter records solely gravity variations between two sites. Galileo invented the unit gal to describe measured gravity. Gravimeters used in geophysical surveys have a resolution of around 0.01 milligal or mgal (1 milligal = 0.001 centimeter per second). That is, they can detect variations in the Earth's gravitational field as tiny as one part in 100,000,000.
- Data processing: Low resolution of the gravity data needs augmentation where data processing plays a vital role.
- Interpretation: To interpret the underlying information a gravity model is built.

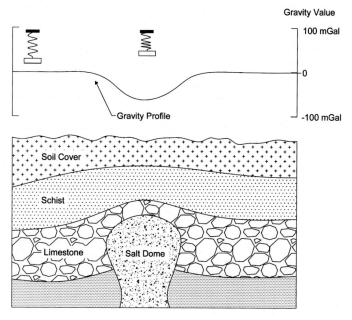

Figure 2.11 In this geological cross-section, the relatively low density of salt with respect to its surroundings renders the salt dome a zone of anomalously low mass. The Earth's gravitational field is perturbed by subsurface mass distributions and the salt dome therefore gives rise to a negative gravity anomaly with respect to surrounding areas, as illustrated in the upper part of the plate by the Bouguer anomaly [45].

4. **How computing can make the survey operation efficient:** Smart solutions and computers, in general, are utilized to improve the quality and acceptability of the survey operations. An example of finding opti-

mized well drilling strategy starting from survey operations is presented in Fig. 2.10. Various forms of computing solutions that help the survey operations are as follows:

- Utilizing drones or autonomous vehicles for survey operation.
- Selecting the optimized path for autonomous vehicles to survey a large area.
- Employing air balloons equipped with Edge systems.
- Real-time processing of the collected data with Edge Computing and predicting or re-estimating the survey path.

2.4.3 Smart data preprocessing on the edge platforms

Sensor-generated data have to be processed before feeding to any analytical model or any automation system. This process is known as data preprocessing. This is a subbranch of data management and mainly includes data cleaning, transformation, and conversion. Typically, smart oil fields generate plenty of unstructured and multi-structured data that need the following preprocessing steps.

2.4.3.1 Data cleaning

Data cleaning includes removing corrupted or noisy data from the generated raw data. It plays a vital role for the efficiency of any analytical model. In particular, the efficiency of any machine learning model depends on the quality of the injected data to the model. For instance, lack of a proper data cleaning strategy in a machine learning or automaton model utilized to identify hydrocarbon extraction in a site can lead to false positive or false negative results that can potentially cause a disaster.

2.4.3.2 Data transformation

Sensor-generated raw data sometimes needed to be transformed from the original format to another format before being fed to the data processing application. This conversion process is referred to as data transformation. For example, in seismic imaging survey generated acoustic data have to be transformed to a 2D/3D image for further analysis.

2.4.3.3 Data conversion

The O&G industry generates a huge amount of raw data that are logged in various formats. For different machine learning or automation models generated data need to be injected in specific format (*e.g.*, text, numeric, video, audio). One example is optimizing the performance of electric submersible pumps (ESPs) [46,47]. Authors in [46], exploited Big Data to analyze the performance of ESPs by identifying emergency situations (*e.g.*, overheating, unsuccessful startups). Within one year period, authors gathered 200 million logs from 1649 wells to perform their study. In this work, collected data were in heterogeneous formats that were converted to comma-separated (a.k.a. csv) file format for further analysis.

2.4.4 Data analytics across Edge-to-Cloud continuum

Data analytics is divided into two types, namely *descriptive* and *predictive*. The descriptive type includes reports, interpretation, and visualization. On the contrary, the predictive type has two categories: simulation; and emulation/physical modeling. For example, seismic processing and reservoir simulation are in the simulation category, whereas, the emulation of oil spill trajectory and the penetration rate of a drilling bit are considered as physical modeling.

2.4.4.1 Descriptive data analytics in the O&G industry

According to a research article [48], descriptive analysis is a process to develop a summary of the historical data which provides the insight and knowledge about what happened. For example, we are able to see, the increase in oil prices after a political news. This analysis provides knowledge about what happened after a particular news. Technically, descriptive analytic is the analysis of data from the past (even if it is from a few seconds ago). It can be the gathering and synthesis of multiple weather instruments to explain the current weather conditions in simple English.

Descriptive analysis results are in form of insights that can be categorically divided into three forms, namely reports, interpretation, and visualization. The reports are generated based on the data that has been collected over a period of time and stored in a data warehouse. Analysis is made by various processes such as data analysts, business analysts, supply chain analysis, and security analysts. The reports are often generated in an on-demand manner and help to make decisions or to understand what has gone wrong in the past and what should be implemented to overcome those problems in the future.

2.4.4.2 Predictive data analytics in the O&G industry

Predictive analytic is essentially the process of using any of one or more statistical techniques to analyze real-time or historical data, with the intent of making some sort of prediction about the future. The subject has become more prominent largely because of growing dialog about a related topic: big data. As business and research entities around the globe produce increasingly complex data sets, tools for not only organizing and storing but also filtering and analyzing the data have become necessary. Predictive analytic is one of the tools that have emerged from this need.

Predictive analysis is used to recognize past patterns to predict the future. For example, some leading companies use predictive analytics for making their score high in the market. Some companies have gone one step further by implementing predictive analysis to analyze the source score, number and type of communications to support marketing sales and other types of complex forecasts. Many data management experts have suggested ways in which predictive analytics can help in the development and management of upstream and downstream facilities.

The upstream process of the oil and gas industry is exploration, development and production of crude oil or natural gas. For upstream, predictive analysis helps companies in asset maintenance improvement, exploration optimization, production optimization, drilling optimization, and risk assessment. For example, based on geographic features data, equipment information and operation team information, a model target can be built to optimize a high rate drilling and production prognosis. Based on this model, companies can easily identify which team process with what kind of equipment can yield the highest work efficiency under specific geographic conditions.

2.4.5 Synergy between High Performance Computing (HPC) and Cloud computing to provide real-time insights from simulations

High Performance Computing (HPC) and Big Data ecosystems are structurally different. The HPC systems are designed for the faster execution of large parallel programs (*e.g.*, rock structure analysis), whereas, Big Data systems are designed to collect and analyze a vast volume of data and handle data-driven applications.

Most scientific simulations are extensive in terms of execution time and generate a large volume of data. The storage, movement, and post-analysis of the generated data are too expensive. So, an efficient way to process the data analysis applications is to continuously process the in-memory data while the simulation programs are running. However, the main problem is that in-situ analysis frameworks and analytic data applications have been designed for two different ecosystems. While the in-situ analysis frameworks are often designed for the HPC ecosystems, most data-driven and machine learning applications have been developed using Big Data programming languages and libraries like Python, Scala, and Spark that lend themselves better to the Cloud ecosystem. Accordingly, one challenge that the scientists and practitioners have dealt with over the past few years is how to design a system that can run scientific workflows consisting of both native HPC and Big Data applications?

Attempts have been undertaken to bridge the divide between these two ecosystems to combine high-performance computing with cloud settings [49]. The objective is to effectively evaluate HPC simulations by using the elastic services and native software available in the Cloud. However, there are numerous obstacles to overcome in order to accomplish the aim. Incompatibility of data formats and bandwidth constraints between HPC and Cloud are two major barriers to merging the two ecosystems. The mapping between simulation and data analysis should be adjusted to avoid data delays. To address these issues, [50] proposes a software architecture called ElasticBroker. ElasticBroker connects the HPC and Cloud ecosystems. When HPC applications based on the Message Passing Interface (MPI) are linked to the ElasticBroker library, the simulation data is transformed into Cloud-native data objects and continuously streamed to data analysis services deployed in Cloud systems, where the data objects are organized and analyzed along with the scheme information. ElasticBroker is composed of two key components:

- A C/C++ brokering library at the HPC end that transforms data from a simulation-domain format to a Cloud-compatible format.
- A distributed stream processing analysis service that is deployed at the Cloud end.

Between HPC and Cloud, data is converted from the simulation by ElasticBroker, and then transferred to the distributed stream processing analysis service using available intersite bandwidth.

2.4.6 Upstream as a cyber-physical system: detection to action framework

The main purpose of employing cyber-physical systems (CPS) in the upstream sector is to achieve automation. The automation management system is primarily divided into two subsystems, namely *detection*, and *decision*.

The **detection subsystem** mainly relies on the machine and deep learning solutions to detect any anomaly that can hamper the smooth operation of smart oil fields. The detection systems utilize sensor-generated data to handle the detection operation. In contrast, a decision subsystem generates long-term or short-term alerts for further analysis of the detected area or equipment. The decision systems utilize actuators for recovery or neutralization of any affected area or equipment. A taxonomy of the detection subsystems is shown in Fig. 2.12. As we can see, the detection systems can be categorized into the "intrusion detection" systems that can be of Physical and Cyber types; and "safety detection" systems that can be for the safety of the environment, human, and equipment. Several instances of each detection system have been developed and some notable mentions are listed in Fig. 2.12.

The **decision subsystem** includes four categories of decision systems, namely Maintenance, Equipment Configuration, Disaster Management, and Data Governance. A brief discussion of these systems is as follows:

- *Maintenance:* The Smart O&G industry encompasses numerous equipment such as valves, drilling equipment, actuators, sensors, regulators, pumps, hose fittings among many others. These equipment need maintenance operations and critical decisions regarding their maintenance that sometimes is crucial for the safety of the entire site. For instance, there are various types of valves to control the liquid or gas flow. These valves must be always accessible and operational in the case of hazards require an immediate action. To avoid jams or malfunction, precise preventive maintenance decisions need to be performed in an efficient way to enable safety of the smart O&G industry. Accordingly, two exemplar decision systems are the valve control system, and the shutdown faulty equipment.
- *Equipment Configuration:* Decisions regarding configuration of various O&G equipment fall under this category. Drilling operation, mud weight adjustment, and drill bit control are examples of such decision systems.
- *Disaster Management:* The smooth execution of the contingency plan of a smart O&G field primarily depends on the decisions made in disastrous situations. The disaster management system includes various subtasks that includes typically notifying mainland, dispatch drones, and sending rescue team to the disastrous area. The synchronization among all these subtasks is crucial and automated decision-based detection systems are developed to handle such complicated workflows.

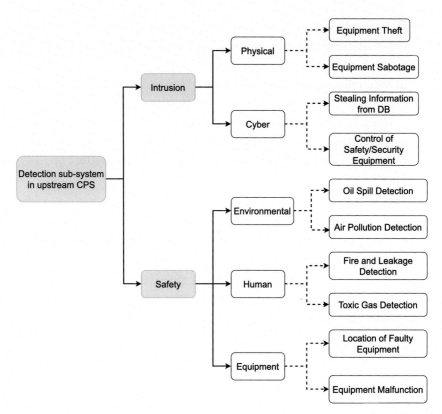

Figure 2.12 Taxonomy of the detection subsystem in a CPS system used for the upstream sector. Example solutions are depicted using dashed-lines and white-boxes.

- *Data Governance (DG):* Data governance is the method of managing the availability, usability, integrity, and security of the data based on the internal data standards and policies that also control data usage within an enterprise system [51]. Effective data governance ensures that data is consistent and trustworthy. A well–designed data governance program typically includes a governance team, a steering committee that acts as the governing body, and a group of data stewards. Data governance along with data stewardship and data quality provide the foundation for properly managing and sustaining information as a key asset in a company.

DG provides a framework of principles, policies, standards, roles and responsibilities, and processes to enable effective information management. It allows the business to take proper and responsible ownership of their information where everyone clearly understands their role and the data they are responsible for. Data governance includes clear descriptions of the tasks and activities that the business must perform to maintain data quality. Data placement decision and reducing downtime of a database server can be considered as two examples of DG in a smart O&G system.

2.4.7 IoT-based monitoring and control solutions

Smart O&G industry vastly relies on the monitoring and control systems for smooth and safe production operation. It utilizes various state-of-the-art technologies, including wired and wireless Supervisory Control And Data Acquisition (SCADA) systems and Internet of Things (IoT). A pictorial view is provided in Fig. 2.13. A brief introduction to these technologies is provided in the remaining parts of this section.

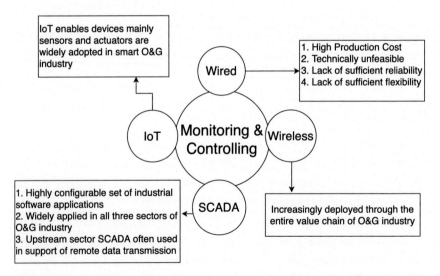

Figure 2.13 Overview of monitoring and control technologies of a smart O&G industry. The significance of the technologies related to monitoring and control are depicted in the rectangle boxes.

2.4.7.1 Wired technology

Various wired communication technologies are used in smart O&G industry that include coaxial cable, optical fiber, and Ethernet cables. Many drawbacks of wired technology make it obsolete for the O&G industry. The key drawbacks of wired communication technology have been reflected in Fig. 2.13. Although, wired communication has many downsides, there are limited areas of the O&G industry that utilize this technology. The list of cable types used in these areas is as follows:

- Copper Communication Cable
- VFD Cable
- Electrical Submersible Pump Cable
- Fiber Optic Communications Cable
- Hazardous Location Armored Cable
- Type E Cable
- Type P Cable
- Industrial Cable

Due to incurring a high cost and flexibility issues, nowadays the utilization of this communication technology is very low. In general, onshore smart oil or gas fields utilize the wired technology more often that the offshore remote ones that mainly depend on the wireless technologies.

2.4.7.2 Wireless technology

Among various wireless communication technology, wireless sensor network (WSN) is widely used in many sectors of the smart O&G industry. The O&G extraction is moving to the remote sites due to scarcity of oil or gas reservoirs in onshore that vastly need wireless communication with the mainland. The two major benefits of the WSN for such environments are cost reduction in cabling and maintenance. WSN is primarily composed of various sensors that is depicted in Fig. 2.14.

2.4.7.3 Supervisory control and data acquisition (SCADA) system

SCADA is a widely used enterprise information technology (IT) system composed of hardware and software entities to enable smooth operations of industrial organizations [52]. An example of cloud-based SCADA system is presented in Fig. 2.15. The main application of SCADA is the industrial management support utilizing customized software and interfaces. SCADA enables specified monitoring of various machines and processes that consist of controllers (*e.g.*, Programmable Logic Controller (PLC) and

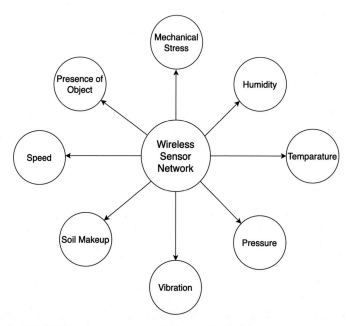

Figure 2.14 Wireless sensor network is composed of various sensors where the sensory activities are utilized to capture or sense data and transmit them to the pertinent destination.

Proportional-Integral-Derivative (PID) controllers), input/output devices, and communication equipment. SCADA usually synchronizes several subsystems with remote equipment that have a limited connectivity. Specifically, SCADA regulates flow, pressure, temperature, and other industrial process variables using its controller units. As such, SCADA systems focus on the following key abilities:

- Data acquisition, preprocessing, and preserving real-time data
- Control and monitoring processes
- Establishing communication between workers and equipment via human-machine interface (HMI) software

The hardware components of SCADA. Among different hardware components, two of the indispensable subsystems are Remote Terminal Units (RTUs) and Programmable Logic Controllers (PLCs). These subsystems are as follows:

RTUs work as data forwarding component. Smart sensors of on-site machines convert physical signal to digital data with the help of the RTUs. Then telemetry hardware is used to send these digital data to the mon-

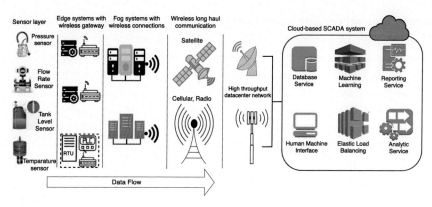

Figure 2.15 The architecture of a SCADA system with the processing and communication layers.

itoring system. Alongside, digital commands from supervisory system are transmitted to the end devices via RTUs.

PLCs are controlling units that monitor many programmed parameters. For example, PLCs perform monitoring of liquid levels, gas meter, voltage current, pressure, temperature and many more. PLCs can be programmed using structured and elementary logic operations. PLCs have significant controlling ability that sometimes replace RTUs working as field devices. To be economical and versatile, PLCs are widely used in the industry, especially in the earlier generations of the SCADA systems. In modern SCADA system PLCs are replaced by industrial PCs that offer more customization ability and a faster execution.

Industrial PCs (IPCs) are used for the process control, data acquisition, and analysis. Unlike PLCs, IPC functions can be updated remotely via cloud-based control systems. This degree of connectivity allows for immediate modification and upgrades to the existing control system. While PLCs were the standard for years, they are not as "smart" as the IPCs that offer advanced features and have the ability to update their programming through the cloud.

Telemetry systems connect PLCs and RTUs to the main control center and data warehouses. They are either wired or wireless. Wired telemetry media in SCADA systems include leased telephone lines and WAN circuits. Their wireless counterparts function based on satellites, radio, and microwaves. Human–machine interface (HMI) presents processed data to the operators. It usually requests data from data acquisition servers.

Computer servers are responsible for the acquisition and management of data for a set of parameters. There are different servers for certain tasks. Some handle alarms, while others file data. Data acquisition servers, for instance, bridge software services with RTUs, PLCs and IPCs via telemetry. **Supervisory system** is a computer or group of computers that handle data during operations and issue commands. A communication infrastructure connects this system to RTUs, PLCs and IPCs.

SCADA software solutions. SCADA shows the real-time performance of equipment. This is achieved via sensors planted on the equipment that collect the data and send the information through Remote Terminal Units (RTUs) and PLCs to the processing unit. Therefore, the SCADA system can pinpoint anomalies by comparing the real-time data against the historically collected data. The software services of a SCADA system are as follows:

Alarm system: This service uses alarms to notify the operator when to take action to resolve a problem. The SCADA system is able to issue alerts with a specific location since it has a network connection with location tracking in most of the sensitive equipment.

Decision making: Permitting maintenance and operational personnel to make more informed and efficient decisions.

Reporting and documentations: The data acquired by the SCADA system are converted to human-readable and documentation that can be presented to and help the management team.

Monitoring: SCADA systems are set up to run commands in critical situations. For example, people set the parameters to 60 percent of the tank's capacity. Any value above that would either alert the teams that need to know about it or let the machine run independently.

Data acquisition: The procedure starts with the RTU, PLC, or IPC. Sensors collect data from numerous pieces of equipment. Each sensor has a specific purpose; for example, some collect data on the flow of water from the reservoir to the water tank. Others, meantime, collect data on water pressure as it escapes the reservoir. After that, the data is assembled and structured such that the control room operator may alter or override standard RTU controls.

Controlling: RTUs, PLCs and IPCs are largely responsible to carry out the control actions. The organizations (especially operators) can intervene in these decisions. For instance, a remote actuator controls the flow of fluids through the pipelines. The SCADA system's software tools, however, allow staff to change the parameters and adjust the alarm conditions.

Data communication: The SCADA systems use wide area network (WAN) and local area network (LAN). A LAN is a group of computers and devices connected to one another, usually within the same premises. A WAN, on the other hand, has a wider reach and connects LANs, thereby allowing the SCADA system to reach equipment in remote locations.

Historian: This software service keeps time-stamped data, Boolean operations, and alerts in a database for later retrieval and processing. By using the Human-Machine Interface (HMI), the SCADA software may query data and create graphs in real-time. This service is often used to obtain the data it requires from a data acquisition server.

Information presentation: HMIs are used in the majority of SCADA systems. RTUs collect and analyze data before sending it to an HMI. The HMI then converts this input into machine-readable signals. It displays a graphical representation of the system and enables users to enter instructions. For example, the HMI may show an image of a pump attached to a certain tank. The operator may monitor the flow and pressure of the water and make changes as needed.

2.4.7.4 IoT in upstream of a smart O&G industry

The Internet of Things (IoT) in the upstream of an O&G industry can be represented as the network of physical objects (sensors and actuators) that are connected to the Internet. Smart O&G industry is getting smarter and more intelligent with the proliferation of IoT in industrial sectors. The main components of the IoT technology that enable a smart solution in the O&G industry are as follows [53]:

- **Smart sensor:** Smart sensors enable the monitoring and detection process that allow the smooth operations of supply chain. In the O&G industry prevention and prediction of any occurrence helps the environment to be protected and improves the efficiency of the production.

- **Smart algorithms:** Smart algorithms perform various validations and critical analysis on the sensor-generated data. To implement the smart solutions, smart algorithms provide valuable perception that enable the control system makes accurate decisions. For example, to avoid uncertain circumstances in the upstream, a smart algorithm makes the decision of starting or stopping the drilling operation.

- **Robots and drones:** Robots and drones are considered as the next generation workers in the hazardous areas of O&G. To implement a smart supply chain, robots and drones being extensively used. For example, robots and drones employed for accurate survey operations in exploration, various maintenance data acquisition, and 3D mapping of landfills.

- **Wearables devices:** The benefits of real-time analytic can be availed by the wearable devices in remote extraction sites to improve the workers' safety. Some examples of the wearable devices are sensor-enable suits, smart watch, smart glasses, and smart helmets. These wearable devices perform continuous monitoring of the workers' physical condition and provide them with real-time advice and/or notification via connecting them uninterruptedly to the control station.

2.4.7.5 Smart IoT applications in the O&G industry

Nowadays, various IoT solutions are evolving for smart O&G industry that transforming the industry with efficient productivity and make it environmentally safe. These smart IoT applications are implemented in various sectors of O&G industry. Among numerous applications of IoT in the O&G industry, some major domain of improvement in the upstream sector can be identified as follows:

Drilling management: Drilling is a major part of oil and gas industry procedures. The Internet of Things proves to be a boon for enhancing efficiency in the drilling procedure. As the rig drills deep, it leads to potentially dangerous circumstances. Rig operators must take precise measurements to extract oil and gas by drilling. Deep-water drilling holds more serious potential for mishaps and dangerous circumstances. IoT devices are beneficial for minimizing risks and carrying out tough operations seamlessly. Smart devices also alert concerned personnel well in advance about any drilling errors using the data received from the sensors.

Offshore monitoring: Most offshore oil and gas production is done in extreme environments. There are very few communication networks available at these rigs. Monitoring temperatures, pressures, and other equipment become a difficult task and an expensive one too. IoT helps overcome these hurdles by providing a truly real-time monitoring system. Using Low-Powered Wide Area Network (LPWAN) [54,55], a lot more monitoring points can be connected. This implementation provides a relatively inexpensive solution for offshore oil and gas rig monitoring. Multiple detectors can be connected to wells within a large area. Each of these detectors can send the data to a central point in real-time. The data can then be leveraged to monitor the drilling and oil and gas production process remotely.

Health and safety: Oil and gas sites are usually found at sensitive and remote locations. The conditions at these sites can be a hazard for the employees working at these locations. IoT solutions provide remote monitoring of equipment and operations, no longer requiring individuals to go to a site without prior knowledge of the situation at hand. Connected sensors and image vision can provide an accurate detail of the situation and help decide the safest course of action. IoT in oil and gas can help to reduce fatalities and injuries caused to employees remarkably. The fatality rates among the oil and gas employees are decreasing, and IoT can help bring down the number to a greater extent. Accidents can prove to be expensive to the companies financially as well as damage the reputation of the company. By using IoT-enabled safety measures, oil and gas companies can provide their workers with a safe working environment. The companies can also benefit from lower insurance and corporate liability.

Carbon footprint control: The implementation of IoT in oil and gas operations results in the efficient functioning of the industry. IoT solutions prove beneficial financially as well as environmentally. With efficient management and working of the field and plant operations, the carbon emissions generated by them can be reduced remarkably. It helps to lower the environmental footprint generated by drilling and production operations that are mainly performed in the upstream sector. Oil and gas companies can, thus, carry out their moral responsibility of not harming the environment while carrying out the O&G operations.

2.4.7.6 Computing industry developing IoT solutions for the O&G

With the emergence of IoT technology, various computing industries are developing solutions for O&G production. The computing industry mainly backed up the IoT technology with the computing and analytical ability in onshore and offshore O&G extraction sites. Solutions from two of the popular computing industry are explained in the following parts.

1. Platform-based IoT solutions for sensor data. Data acquisition with smart sensors utilizing IoT technologies allows the solution providers to make quicker and faster decisions in the O&G industry. Smart sensors can send the data to the processing unit that helps the control unit take necessary actions for the mission-critical process. For example, the hydrocarbon flow from a production well faces obstacles with the increase in its maturity due to the accumulation of liquids. This significantly reduces the production rate, which can be resolved by predicting when the well requires maintenance. In case of de-liquifying the well, the solution can be plunger-lift technology. Intel addresses this problem with faster data acquisition to increase the production rate [56]. Fig. 2.16 represents the plunger

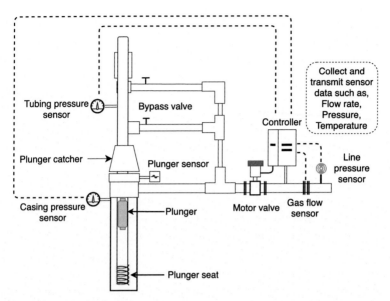

Figure 2.16 Implementation of IoT for data analytic in plunger lift system can improve the well oil production. This data analytic solution provided by Intel for O&G industry.

lift system with a controller that transmits the sensor-generated data for further processing. According to Fig. 2.16, the plunger is sent from top to sit in the plunger-seat located in the bottom hole. With each sales line, liquids collect in the wellbore, increasing backpressure and slowing gas production. As a result, the high pressure lifts the plunger that clears the accumulated liquid restoring the normal production environment.

The IoT technology enables the uninterrupted connection with all the wells' sensors, acquiring sensor data in a continuous manner (*e.g.*, every 30 seconds). Hence, edge computing can analyze sensor data that helps optimize the plunger lift cycle (a plunger dropping from the top to the bottom hole, and with pressure buildup, it goes back to the top, clearing the accumulated liquid). The outcome of this optimization is the improvement in the production up to 30%, which is reflected in the Intel report. Hence, the gathered data can be utilized to rank the wells according to their efficiency and maintain an effective asset management profile. Some of the solutions supporting computing platforms that have a significant impact in the O&G industry are as follows:

- *Image efficiency and simulations platform:* Oil and gas companies use accurate and precise simulations to research old oil fields. The simulation is used from reservoir research to the end of production and sometimes during the well's life cycle. The simulation needs to be done as well as possible and try to make the most money. So, every part of the optimization process and the simulation needs high-performance computing hardware to run well. Fabric bottlenecks and storage walls can make it hard to see how much computing power can help. Special hardware is needed to run this simulation. Intel SSF, Intel Xeon processor E5-2600 v4, and Intel Xeon Phi processors are some of the things that can help.

- *Reservoir visualization:* The quality of the images and how they look are critical to the success of oil and gas exploration. For complex algorithms and processes that need many resources to run, unique computing resources (like Intel Xeon Phi processors) are used as the processing backbone of reservoir visualization applications. This means that visualization applications in the oil and gas industry need a High Performance Computing (HPC) infrastructure with resources that are specific to that industry.

- *High-performance computing for rapid discovery:* The massive amount of data from different oil and gas exploration surveys needs high-performance computing resources, like Intel Scalable System Framework (Intel SSF) and analyzing software tools to search through the data and find valuable information in a short amount of time.

2. Machine learning-based IoT solutions from sensor data. Industrial IoT (IIoT) bridges the gap between industrial equipment (a.k.a. Operations Technology (OT)) and automation networks (a.k.a. Information Technology (IT)). In IT, the use of machine learning, cloud, mobile, and edge computing has become a commonplace. IIoT brings machines, cloud computing, analytics, and people together to improve performance, productivity, and efficiency of industrial processes, thereby, enabling industry leaders to have applications for predictive maintenance and remote monitoring of their operations.

In the case of Amazon cloud, the AWS IoT services can be leveraged by different industries (such as mining, energy, and utilities, manufacturing, precision agriculture, and oil and gas) to reason on top of the operational data, hence, improving the performance, productivity, and efficiency. Below, we explain some prominent applications of AWS cloud services for machine and deep learning in the O&G industry:

1) DeepLense for compliance: AWS DeepLense is a video camera with deep learning ability that can run deep learning methods against the captured

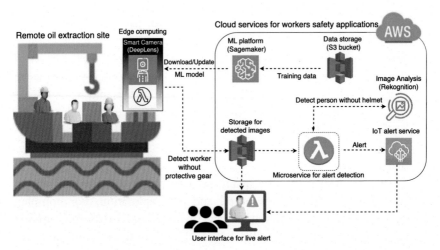

Figure 2.17 Workers safety in remote oil extraction site utilizing DeepLens and Amazon Rekognition.

images in a real-time manner. Utilizing DeepLens, developers of different industry fields can deploy various image processing activities in their own contexts. The main difference between DeepLens and other AI-powered camera lies in the capability that makes it possible to run deep learning inference models locally without sending the video frames to the cloud. Therefore, DeepLens provides an effective platform to address problems occur in places like oil fields (Fig. 2.17) that are often located in remote areas with low or no access to the cloud. As one use case, we can consider the case of assuring the safety compliance of workers in a remote oil field by wearing the protective gears (*i.e.*, helmets and goggles). As shown in Fig. 2.17, DeepLens can be utilized and configured to notify the compliance team about worker who do not follow the safety regulations. The DeepLense installed on a remote oil extraction site, download the compliance detection ML trained model from cloud services. Hence, the AWS lambda function [57] is utilized to perform the inference operation within the DeepLense camera using its local GPU. The DeepLense camera captures video frames used as the inference operation's input. As depicted in Fig. 2.17, the output of the lambda function is sent back to cloud services to store (S3 bucket) images and generate alerts (a.k.a. micro-service) to appropriate personnel. Finally, the compliance monitoring team gets the live alert with the worker image not following compliance guideline. Therefore, DeepLense ensures workers' safety in remote offshore oil extraction sites.

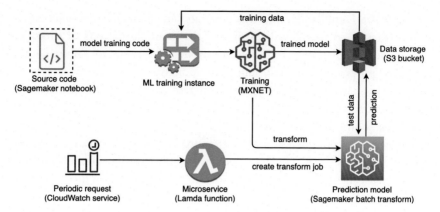

Figure 2.18 Predictive maintenance using machine learning architecture on AWS. The solution uses an ml.t2.medium Amazon SageMaker notebook instance to orchestrate the model, but it uses an ml.3p.2xlarge Amazon Sage-Maker training instance to perform the training. The training code and trained model are stored in the solution's Amazon S3 bucket.

2) Predictive maintenance: Industry bears a colossal weight as a result of equipment failure. It is estimated that unscheduled downtime affects productive plant capacity by 5 to 20% and costs industrial firms $50 billion per year [58]. The cost of repairing or replacing equipment may be high, but the ultimate cost of unforeseen equipment failure is the repercussions. Moreover, the impacts of failing equipment reverberate through interrupted upstream activities, increasing exposure to safety concerns and additional losses. Predictive Maintenance analytics mainly focus on identifying potential failures before any production process is interrupted. One of the solutions from Amazon cloud in the field of predictive maintenance is the identification of potential equipment failure using the machine learning (ML) model [59]. The Fig. 2.18 represents a workflow architecture of AWS machine learning services for the predictive maintenance use case. According to Fig. 2.18, a SageMaker (a fully managed to machine learning service to prepare, build, train, and deploy high-quality ML models) notebook instance is used to orchestrate the model, and a SageMaker training instance is used to train the model. Both the training code and the trained model are kept in the Amazon storage (S3 bucket) associated with the solution. In addition, the solution installs an "Amazon CloudWatch Events" rule that is scheduled to execute every day. The solution is set up to apply the trained model and forecast the remaining usable life (RUL) from the example dataset to activate an AWS Lambda function that generates an Amazon SageMaker batch transform task. Amazon utilizes one of the example datasets of turbofan degradation simulation to train the ML model for equipment failure detection. The purpose is to rapidly suggest a predictive model that, by analyzing sensor data characterizing the engine's current and historical performance, can anticipate the number of time-steps remaining until the equipment fails. This solution can be implemented in the O&G industry to automate the detection of potential equipment failure, including required actions to trigger upon detection. The ML model is flexible to work with any custom dataset that can fit in various sectors of the O&G industry. Fig. 2.18 illustrates the architecture that may be created by referencing the sample code on GitHub [60]. The code provides an instance dataset of a turbofan deterioration simulation in an Amazon Simple Storage Service (Amazon S3) bucket, as well as an Amazon SageMaker endpoint with a machine learning model trained on the dataset to forecast the RUL of a turbofan.

2.5 Summary

The highly competitive nature of O&G industry has been exacerbated by the uncertainties surrounding global demand for oil and natural gas. This has increased the demand for businesses to minimize their operational and capital expenditures. Simultaneously, the sector is experiencing a digital revolution due to the advent of the Internet of Things (IoT) and technologies like edge-cloud computing, wireless networks, that are combined with machine learning. The O&G industry can take advantage of this development to enhance productivity and minimize expenses, allowing it to compete in today's market. With the emerging solutions from IoT, the O&G industry is becoming robust in many fault-intolerant sectors (*e.g.*, drilling, and oil and gas extraction from wells). In addition, failures in these fault-intolerant activities can lead to disasters that pollute the environment drastically. It has been a challenge for the O&G industry to protect the surrounding environment, and keep the work place safe for the on-site workers by making the idea of unmanned sites a reality. In this chapter, we explored various resolutions utilizing state-of-the-art IoT, edge-cloud, and Machine Learning technologies to keep the environment clean, improve the efficiency of the upstream O&G industry, and help it to become safe and ultimately unmanned. In the next section, we will focus on the midstream industry and the use of smart solutions in that context.

CHAPTER 3

Smart midstream of O&G industry
Smartness in midstream sector

3.1 Introduction and overview

Midstream is the oil and gas sector activity connecting producing regions to population centers with industrial and residential consumers. Midstream activities are primarily concerned with the storage, transportation, and processing of crude oil or natural gas. Midstream, in general, links upstream and downstream. Typically, midstream operations begin after wellhead treatment and conclude when a hydrocarbon stream reaches refinery storage tanks (*i.e.*, the tank farm). Natural gas liquefaction [61] to LNG, bitumen extraction/dilution, and the creation of synthetic crude oil (syncrude) for transportation all belong in the midstream rather than upstream sector due to the extensive downstream skills required; reference [62]. The midstream sector's systematic taxonomy is represented in Fig. 3.1. As a result, the midstream industry may be broadly classified into four primary subsectors: transportation, storage, safety, and security. According to Fig. 3.1, the transportation subsector is further divided into three categories, namely short, medium, and long distance, respectively. The storage subsector is divided into three categories that are open top, fixed roof, and floating roof. Then the safety subsector is categorized into two categories: workers' health and environment. Finally, the security subsector is categorized into cyber and physical categories. Additionally, to represent examples of each category, white boxes are depicted at the bottom of Fig. 3.1. To complement this midstream taxonomy, we investigated the computing characteristics of these four subsectors and created a new taxonomy of computing for midstream. This chapter is chronologically divided according to these four subsectors.

3.1.1 The use of computation in midstream

The emerging IoT solutions and advancement in computing technologies have significantly changed the midstream sector's operations. Along with improvements in the transportation and storage subsector processes in midstream, computation plays a vital role in ensuring midstream safety.

IoT for Smart Operations in the Oil and Gas Industry
https://doi.org/10.1016/B978-0-32-391151-1.00012-5

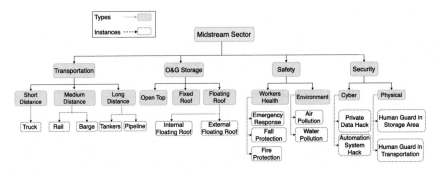

Figure 3.1 Taxonomy of the midstream sector for the O&G industry is presented. Four main subsectors are identified where various midstream operations take place.

Therefore, a well-defined taxonomy for identifying the computation involvement in midstream is necessary for the researchers and scholars to address the scope of computation. As such, Fig. 3.2 represents the computing taxonomy of the midstream sector of the O&G industry.

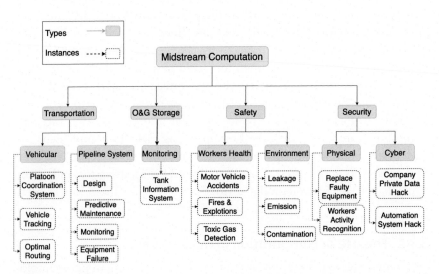

Figure 3.2 Taxonomy of the use of computation in midstream sector. Solid lines in the taxonomy define concept categories whereas dotted lines define examples.

3.2 Transportation of O&G

The Midstream industry is responsible for linking remote oil or gas producing areas and distribution centers where consumers are located. Depending on the hydrocarbon extraction site's distance and accessibility to the mainland, various forms of transportation (*i.e.*, truck, rail, barge) are utilized to transfer the producing oil or gas. Hence, considering the distance between production and distribution centers, transportation can be divided into three categories, namely, short, medium, and long distance transport, which is represented in Fig. 3.3. In addition, depending on the location of the extraction sites, the midstream transportation can be divided into two zones, namely *onshore* and *offshore*. In oil and gas transportation, pipelines account for a significant share of the transport medium. There are different types of pipelines used in the oil and gas industry. Various criteria, including the 'product' being carried, the stage of delivery, and whether the pipeline is utilized in the upstream, middle, or downstream sector, are used to determine the pipeline type. Thus, Fig. 3.3 provides a more comprehensive perspective of the midstream transportation business by factoring in distance and location considerations from extraction sites to distribution hubs.

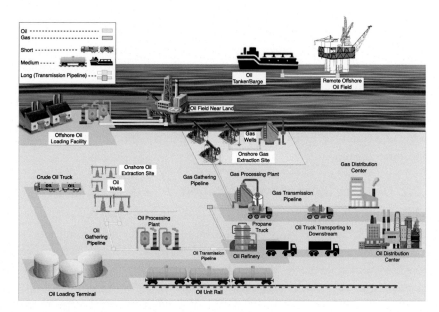

Figure 3.3 Midstream transportation in terms of distance to travel from extraction site to downstream.

3.2.1 Various types of pipeline

Pipelines are typically used in long distance transportation. Although, Each type of pipeline in the O&G industry is suitable for performing an activity. Different types of pipelines are categorized in the following manner:

- Gathering Pipeline: These pipes are called "gathering pipelines," and they help move oil or gas from the source to the processing plants or storage tanks. These are usually fed by "Flowlines," which are pipes that connect to wells in the ground. In addition, subsea pipes used to collect the product from deepwater production platforms are also in this group.
- Transmission Pipeline: Transmission pipelines are used for long distance transport of crude oil, NGLs, natural gas, and refined products across states, countries, and continents.
- Distribution Pipeline: Distribution companies use 'Mains' and service lines to ensure that people get the water they need. In addition, they work together to bring natural gas to the neighborhoods where homes and businesses are built.
- Flowlines: Flowlines merge to a single wellhead in an operational oil-field. Their mission is to transport raw material from the wellhead to the collecting lines. They are no larger than 12" in diameter and transport a combination of oil, gas, water, and sand.
- Feeder Pipelines: Feeder pipelines transfer the product from processing facilities and storage tanks to the long distance transmission pipelines. The product might be crude oil, natural gas, or natural gas liquids. Feeder lines are typically 6 to 12" in diameter.

Considering the variety of pipeline systems in O&G industry, it is essential to identify the usage scope of different pipelines along with extraction location and distance. Table 3.1 represents various transportation types along with different pipeline systems.

3.2.2 Distance issue of transportation

3.2.2.1 Short distance transportation

Short distance transportation in midstream mainly refers to the transportation of oil in a storage truck (Fig. 3.3 represents an example of a storage vehicle). Even while trucks have the lowest storage capacity of any oil delivery technique, they can also go to most places with considerable flexibility. Moreover, according to [63], trucks are also often the final stage in the transportation process, bringing crude oil and refined petroleum products to their appropriate storage locations.

Table 3.1 Categorical Transportation with respect to distance and location type. Multiple transport mediums are utilized for different transportation conditions and depending on the transporting materials (*e.g.*, crude oil, natural gas).

	Short Distance	Medium Distance	Long Distance
Onshore	Truck, Flowlines Pipeline, Rail	Rail, Gathering Pipeline	Transmission Pipeline
Offshore	Flowline Pipeline, Ship	Ship, Barge	Transmission Pipeline, Tankers

3.2.2.2 Medium distance transportation

Rail cars and barges are mostly utilized in medium distance transportation that is depicted in Fig. 3.3. Oil shipment by train has become a growing phenomenon as new oil reservoirs are identified across the globe. The relatively small capital costs and construction period make rail transport an ideal alternative to pipelines for long distance shipping. However, speed, carbon emissions, and accidents are some significant drawbacks to rail transport [63]. On the contrary, when land transportation is not feasible, barges or ships are considered the primary vehicle of transportation. A typical 30,000-barrel tank barge can carry the equivalent of 45 rail cars at about one-third the cost. Compared to a pipeline, barges are cheaper by 20–35%, depending on the route. Tank barges traditionally carry petrochemicals and natural gas feedstocks to chemical plants. The drawbacks are typically the speed and the environmental concerns [63].

3.2.2.3 Long distance transportation

Pipelines and tankers are the primary modes of long distance transportation (Fig. 3.3). Pipelines transmit large amounts of crude oil, either above or under land or underwater. Pipelines transport crude oil from the wellhead to gathering and processing plants, refineries, and tanker loading terminals. According to truck transport, pipelines use much less energy than trucks or rail and reduce carbon impact. However, excess water must be removed from the pipeline during crude oil transit to fulfill standards. According to [62], the maximum permissible water content for pipeline transportation is between 0.5 and 2.0 percent.

3.2.3 Challenges of transportation systems

Other than pipelines, other transportation mediums (*i.e.*, trucks, rail cars, barges, and tankers) have temporary storage where oil or gas is stored for a short period until entering refinery storage tanks. In temporary storage, raw oil or gas must maintain certain pressure, temperature, and concentration level to avoid any incident. Hence, ensuring the safe transfer of oil and gas from the production site to the transportation medium and keeping them in temporary storage is one of the primary challenges in the transportation field.

Considering a safe environment while producing oil and gas, leakage from transportation fields can pollute the environment significantly. For example, oil tankers and barges for sea transportation can cause natural disasters like oil spills and contamination of harmful chemicals in the water. Similarly, leakage of toxic gas from the transportation field can pollute the environment drastically.

One of the biggest challenges in the midstream sector is the cost-efficient transportation of oil and gas. Due to the scarcity of potential reservoirs in mainlands, the O&G industry has to expand the extraction operations in remote areas located in distant places. Hence, cost-efficient long distance transportation is a challenging problem in the midstream sector. One way to reduce the cost of transportation is to find the best path or route of transportation. In all forms of transportation, pipelines are connected explicitly or implicitly.

Additionally, the pipeline is the most efficient medium of transportation that significantly impacts the transportation subsector of midstream. Considering pipelines as the primary means of transportation, they are prone to leakage accidents that reduce the industry's profit and pollute the surrounding environment in extreme situations. To that end, addressing the leakage detection systems for transportation medium helps build a robust transport system in midstream. Therefore, various leakage detection systems are categorized in the following sections to understand the solution space better.

3.2.4 Leakage detection systems in O&G transportation

The utilization of pipeline is considered a primary means of transporting extracted crude oil or gas [64]. Any means of transportation in Oil and Gas is somehow connected to the pipeline, whether it is onshore or offshore. Although pipelines are the most reliable and safe option compared to other transportation methods, accidents and thefts can occur with pipelines. As such, leak detection systems can avoid damage to people, the environment, and the high costs for repair, renovation, indemnity, breakdowns, and the lost value of the liquid or gas that has been released. Various forms of the leakage detection system are categorized in Table 3.2 considering the transportation mediums (*i.e.*, truck, barge, pipeline).

From Table 3.2, it is clearly visible that the pipeline has a major leakage problem. Therefore, many leakage detection systems have been developed to identify the source of leakage and mitigate the problem as soon as possible. In most solutions, computation is used extensively, either in implementing the solution or improving the existing one.

3.3 Smart transportation

Advance computation technologies have transformed the midstream sector of the oil and gas industry by changing manual operations to automation systems. Following the taxonomy in Fig. 3.4, computation in transportation can be categorized into two categories, namely vehicular and pipeline systems. The main purpose of utilizing computation or digitization of vehicular subsector of transportation is to help or improve oil and gas transportation from the production site.

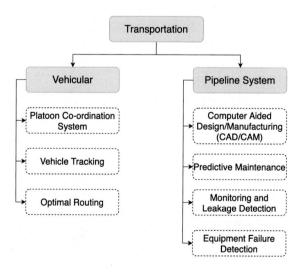

Figure 3.4 Taxonomy of computation in transportation of midstream sector. The main two branches are colored, and white background represents the instances or examples of related branch.

Table 3.2 Categorical leakage detection systems applicable for transportation media.

Transportation	Exterior Methods				Visual Methods			Interior Methods		
Leakage Detection Systems in Transportation of Midstream	Smart Pigging	Sensors	Fiber Optic Sensing	Acoustic Sensing	AUV/ Drones	Helicopters	Trained Dogs/ Human	Mass/ Volume	Pressure Point Analysis	Dynamic Modeling
Truck		✓			✓				✓	✓
Rail (Tank Wagon)		✓			✓				✓	✓
Barge		✓			✓				✓	✓
Tanker		✓			✓	✓	✓		✓	✓
Pipeline	✓	✓	✓	✓	✓			✓	✓	✓

3.3.1 Smart vehicular transportation

Vehicular transportation mainly refers to the transfer of raw oil and gas via any vehicle (*i.e.*, truck, railway, ship) from midstream to downstream. Due to the sensitive nature of the unrefined oil and gas, vehicular transportation has to consider various aspects of the transferring vehicle and the route or path it chooses to travel. Additionally, maintaining proper pressure, temperature, and insulation in temporary transport tanks of transportation vehicles need to be controlled and monitored where computation can aid in developing robust systems. Hence, considering the significance of computation in vehicular transportation, some prior approaches are discussed in the following section.

3.3.1.1 Platoon coordination system

Following transportation definitions, platooning is a system for driving a group of vehicles together. The platoon coordination system mainly increases the capacity of the road by utilizing an automated highway system [65,66]. Advancements in information and communications technology, as well as in onboard technology, allow the vehicles to connect, share the information [67], and the infrastructure. Fig. 3.5 represents a platoon system where information is shared from left to right (*i.e.*, from platoon leader to tail member). According to Fig. 3.5, a platoon leader stays in front of a platoon and guides the platoon to reach its destination safely. The platoon coordination-related information (*e.g.*, speed, gap, vehicle health) is shared from the leader to the platoon members. During the oil and gas transfer with an autonomous platoon of vehicles, privacy and security are essential factors. While information sharing, mainly a collective machine learning process, utilizes various privacy protocols in the platoon coordination system. In [68], Fu et al. explored the existing security aspects of connected autonomous vehicles and proposed a blockchain-based collective learning method for the improvement of the system.

3.3.1.2 Vehicle tracking

Vehicle tracking systems mainly denote the monitoring of the vehicles from their source to destination for supporting any emergency purposes as depicted in Fig. 3.6. Vehicle tracking plays a vital role in avoiding unwanted accidents. In addition, with vehicle tracking systems, it is possible to stop or recover stolen oil or gas while transferring. Advanced vehicle (GPS tracking) tracking technologies may assist enhance the oil and gas

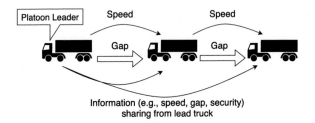

Figure 3.5 An example of platoon coordination system that maintains secure gap and speed among the platoon leader and followers throughout the trip.

fleet's operations in various ways. The system, for example, may immediately check the whereabouts of vehicles and assets, acquire on–demand access to vast volumes of data, and assist in the launch of a green initiative, as shown in Fig. 3.6. With a map view the oil and gas sector may monitor fleet operations at any time and from any location. It is critical to know what is going on with the fleet in real-time if you want to keep it running well. The O&G business may see fleet activities in various ways, examine updated maps, and connect with users/drivers for dispatching and messaging reasons using an advanced solution. Some oil and gas businesses work in distant locations where coverage is problematic. Advanced GPS tracking technologies allow the supervisor to mix and match cellular providers to optimum coverage. GPS tracking devices "store forward" when there is no cellular service. The satellites monitor the vehicle or asset as it leaves the coverage area and gathers data. When it regains cell service, all of the data is sent to the GPS tracking provider's servers for seamless reporting as if they had never lost service. The O&G executives may also use real-time location information to swiftly check the whereabouts of vehicles and nonpowered equipment to safeguard the fleet from theft. Nonpowered assets, such as generators, pumps, and excavators, are often attacked. It might be almost hard to identify and retrieve stolen items without GPS monitoring. As a result, a GPS monitoring system helps safeguard the fleet against theft.

3.3.1.3 Optimal routing

Typically oil or gas production sites are located in remote areas or in the sea. Hence, transferring extracted raw petroleum products to the downstream facilities is challenging due to various rough and uncertain routes. Optimal routing is an intuitive solution that can save money and time. Traditional route design has produced delivery issues that have impacted the petroleum product distribution system. The customer, a renowned petroleum firm in

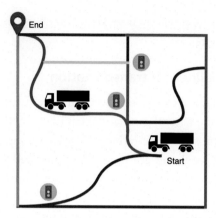

Figure 3.6 An example of vehicle tracking system that keep track of the vehicle from start to end. It broadcast the route and traffic condition and track the vehicle for emergency support.

the United States, was experiencing difficult issues [69] as a result of the high cost of gasoline and the ever-increasing need for dependable delivery services. They needed to improve the efficiency with which they planned their transportation timetables and routes. Furthermore, they wanted to satisfy strict delivery deadlines, offer prompt customer support, and give instant information on any delivery route deviations. The customer sought to cut transportation costs and keep track of the movement of pallets, items, or deliveries more easily with the use of route optimization technologies.

Finding the most direct and cost-effective routes by taking into account all of the potential rest stops and destinations is referred to as Route Optimization. Processes are simplified, fuel costs are reduced, and overall efficiency is improved. A shipping company's performance depends heavily on its ability to optimize its routes. According to a recent analysis of the midstream sector, researchers have found that route optimization is critical to the petroleum business.

Route optimization [70] technologies may address several delivery and distribution issues. The solutions help the client minimize route planning time, better communicate with customers about orders, and satisfy other customer needs. By utilizing various advanced analytics (*e.g.*, cluster analysis, conjoint analysis, scenario analysis, and what-if analysis) on the historical and real-time route and weather information, the solution application can suggest the best optimal route for oil and gas transportation. The solutions may also help them improve their services and check the financial sustainability of a daily route.

3.3.2 Smart pipelines for transportation

Pipelines are considered one of the safest, most efficient, and economical means of transporting natural gas, natural gas liquids (NGLs), and crude oil from the production site to the refinery. Midstream pipelines are generally constructed out of steel and are buried underground, typically at depths of at least three feet. Special coatings and cathodic protection, which employ small electrical currents applied to the steel tube, are utilized to guard against corrosion. From placement to start transferring extracted oil and gas, the pipeline has many applications that need to be maintained as a whole and termed as *pipeline system*. Among several operations of pipeline systems, we will discuss four significantly important operations that need computing aid to make the system robust against any uncertain incidents. The four important computer-aided operations are pipeline placement design, predictive maintenance, pipeline monitoring, and installed equipment failure detection.

3.3.2.1 Pipeline design

Pipeline system design needs a lot of calculations and measurements. For this reason, computer-aided simulations are helpful for pipeline system design and implementation [71]. The primary purpose of the simulation model is to guide developers to create a system that simulates the dynamic behavior of a pipeline network system with different system configurations and varying diameters, lengths of pipes, size of compressors, and total natural gas consumption [71]. In addition, optimizing parameters for the pipeline system is a challenging problem in the pipeline design field. In [72], authors have proposed a genetic algorithm for optimizing system parameters for pipeline systems.

The technological advancement has significant improvement in designing pipeline facilities that require accurate and precise measurement [73]. Specifically, software tools and models (*e.g.*, advanced GPS system for survey, 3D terrain models in hydraulic modeling, stress analysis, and earthwork modeling) help the design engineers perform various design projects with great speed and accuracy.

3.3.2.2 Predictive maintenance

The Midstream sector leverages IIoT to ensure the safety and reliability of the piping, crude oil treatment systems, and gas treating equipment [74]. For example, fiber-optic distributed acoustic sensors, ultrasonic sensors, and temperature sensing systems detect sound variations signaling liquid (*e.g.*, crude oil) leakages, and hydrocarbon sensing cables are used to detect hydrocarbon leaks.

The data from sensors is combined with contextual data (*e.g.*, the data from export facilities, geolocation, weather data) and analyzed against predictive models. Once an abnormal deviation in sensor readings is detected, an IIoT solution triggers an alert, notifying maintenance specialists of a pipeline malfunction. In the midstream, SCADA and plant historians are critical for monitoring real-time process data, but tapping into data from electronic flow meters (EFMs) and keeping track of measurement and flow data is also vital. Because significant transfer actions are conducted in the middle of the process, even a minor measurement inaccuracy might cost a lot of money. As a result, monitoring and analyzing measurement data in the middle of the process is just as important as monitoring and analyzing real-time process data.

Midstream companies employ measurement software for government reporting and compliance with API 21.1 for gas measurement and 21.2 for liquid measurements, according to [39]. Customers of midstream firms frequently request statistics that demonstrate how much gas they are consuming every month. Operators and control room engineers must hustle to construct such reports by analyzing EFM data without measuring tools that can swiftly generate consumption reports. Measurement software can also maintain track of lost and unaccounted products, which is essential for determining the balance of products entering and leaving pipelines and other assets. The entire system balance across all pipelines and their meters or measures calibration tracking is critical for midstream to be up and running efficiently. As a result, predictive maintenance solutions that can detect variances swiftly recalculate volumes and turn data into information.

3.3.2.3 Smart pipeline monitoring

Pipelines are subjected to various issues such as pipeline leakage, corrosion, cracks, flow, and vibration [75]. The oil and gas pipeline industry is focused on enhancing the functionalities of pipelines. Transformational trends, such as operational intelligence, infrastructure development, demand and supply, convergence, greener concepts, and new technologies drive the need for enhanced and innovative sensing technologies in the oil and gas pipeline industry. The industry needs to identify the critical elements of faster transformation and devise strategies that accelerate the adoption of novel sensing platforms.

Safety assessment of oil and gas (O&G) pipelines is necessary to prevent unwanted events that may cause catastrophic accidents and heavy financial losses. In [76], the researchers developed a safety assessment model for O&G pipeline failure by incorporating fuzzy logic into the Bayesian belief network. The proposed fuzzy Bayesian belief network (FBBN) explicitly represents dependencies of events, updating probabilities, and representation of uncertain knowledge (such as randomness, vagueness, and ignorance). The study highlights the utility of FBBN in the safety analysis of O&G pipeline because of its flexible structure that fits a wide variety of accident scenarios. Furthermore, the sensitivity analysis of the proposed model indicates that construction defects, overload, mechanical damage, bad installation, and quality of workers are the most significant causes of the O&G pipeline failures. The research results can help owners of transmission and distribution pipeline companies and professionals to prepare preventive safety measures and allocate proper resources.

3.3.2.4 Equipment failure and leakage detection in pipeline system

The pipeline is considered one of the safest and quickest ways to transport hazardous substances from one location to another. However, pipelines are prone to structural failures due to the continuous influence of material flows and unpredictable environmental conditions. As a result, corrosion, cracks, leakage, and debonding create fatal damage to the pipeline systems. Among these issues, only corrosion contributes to 61% of the entire leakage problems in pipelines [77]. Therefore, monitoring and detecting pipeline failure is essential as it becomes a global issue for O&G industry.

3.3.2.5 Smart management system for long distance pipeline

Big Data and Cloud Computing can be utilized to automatically detect a leakage accident in a pipeline system, localize the point of leakage, and measure the amount and rate of leakage. These technologies are not only useful to detect and localize the leak accident early, but they can also help make the best decision in emergencies.

In [78], He et al. propose a smart pipeline system that can be used to manage leakages in a pipeline. In this work, the fluid flow in pipeline is considered a one-dimensional flow that satisfies the conservation of mass, momentum, and energy. As shown in Fig. 3.7, these equations are employed by different modules to detect and localize the leakage in the pipeline. In a leakage accident, the leakage point is determined by transmitting the pressure and other flow data to the leakage localization module. Then, based on the flow characteristics and data collected by ASPEN database, the leakage rate is calculated by another module. The shape and size of the leaking orifice is also predicted at this step. Finally, the pipeline emergency response module is responsible for actuating the valves along the pipeline to minimize the leakage in the shutdown process of the pipeline.

As shown in Fig. 3.7, the long distance smart pipeline management system integrates the whole lifecycle data of the pipeline through the architecture of terminal, Cloud, and Big Data that provides intelligent analysis and decision support by Cloud Computing and distributed computing. At the same time, the system also improves the quality, rate of progress, and safety management by means of information, which is to realize the visualization, networking, and intelligent management of pipelines. Finally, a safe and efficient smart oil and gas pipeline network with the capability of comprehensive perception, automatic prejudgment, intelligent optimization, and self-adjustment is formed. Furthermore, it will significantly promote the technology level of pipeline integrity management by enhancing pipeline data acquisition, risk assessment, prevention and control, defect detection, and evaluation.

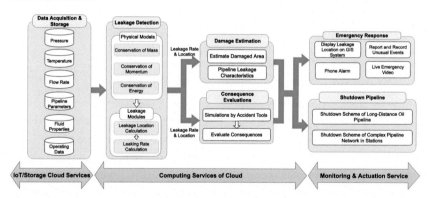

Figure 3.7 Overall design—the technical architecture of the long distance pipeline smart management system [79].

Cloud Computing, Big Data, and other technologies are used as core modules to store, calculate, process, and analyze pipeline data collected by the perceptual layer. Then, the real-time control, integrity management, and scientific decision-making of long distance pipelines can be realized. In this technical framework, Cloud Computing technology can be applied to integrate and store massive, real-time, and accurate data from inductive components embedded in oil and gas pipelines, equipment and facilities, environmentally sensitive points, and maintenance tools. Using Big Data technology, one can dynamically analyze data from multiple perspectives and multi-levels, explore the potential value of data with specific businesses, optimize the process, and prevent and control risks to ensure pipelines' safe and efficient operation. With the IoT, all data of the long distance pipelines can be transmitted in real-time through field buses, industrial Ethernet, and satellite communication.

3.3.2.6 Equipment failure detection

As shown in Table 3.3, pipeline systems consist of several pieces of equipment (*e.g.*, initial injection station, compressor/pump stations, partial delivery station) that combine together to transport crude oil or gas from upstream to downstream. Nowadays, various sensors are implemented into the pipeline system to track the abnormalities or deviations in real-time and report instantly to the control room. For instance, a pressure drop from any pipeline section represents a potential leak. Furthermore, sensors can help detect structural issues of pipelines in the very early stages that may lead to a severe spill or fatal explosion. For example, an initial crack inside the pipeline can be identified using ultrasonic and acoustic sensors. Similarly, corrosion of the pipeline system can is detected with magnetic sensors that measure the change in pipeline wall thickness [79].

3.3.2.7 Case study: pump failure cause investigation

Pumps are fundamental units in pipeline systems that help transport oil and gas substances from production sites to downstream. The maintenance costs of industrial pumps are 7% of overall equipment maintenance cost, while pump failures are responsible for 0.2% of production loss. This fraction of loss seems to be small compared to the total loss, whereas its impact considering the amount of money is huge. Hence, identifying the cause of pump failure or analysis of the root cause for pump failure is significantly important for reducing loss in production.

The general practice for most pumps is maximizing their lifetime. However, the optimal run life of an electrical submersible pump is five years, compared to the average run life of only 1.2 years. This is a major issue, since electrical submersible pumps are used for 60 percent of oil and gas

Table 3.3 The equipment of pipeline system and their usage is presented in a tabular format.

Pipeline Equipment	Usage	Failure Detection Technique
Pumps and Compressors	Move hazardous liquid and natural gas through pipelines	Machine Learning (SVM) for Pump Cavitation Detection [80]
Metering Equipment	Measure the amount of product being received or delivered	Machine Learning (Isolation Forest) Method [81]
Remote or manually operated block and control valves	Control the flow of moving liquid	Acoustic Emission Parameters and Support Vector Machine Approach [82]
Relief valves and other overpressure control devices	Prevent rupture of the pipeline due to unexpected pressure surges	Predictive Maintenance
Tanks	Store hazardous liquids. Equipped with level gauges that warn operators that the tank is near its maximum capacity.	Tank leak detection with sensors

production. While a pump failure in a refinery may only affect one part of a process, pump failures in an oil field can shut down a well or pipeline.

So, what causes that 3+ years of operational lifetime discrepancy? It is very easy to say nearly all the faults lie with user error. But there are so many things that can go wrong. Often, centrifugal and positive-displacement pumps are treated the same, so the fixes will actually cause problems over time. Pumps will operate at improper speeds, be misaligned, or simply be the wrong size (but still installed), and there can always exist contamination in the lubricant. Fig. 3.8 shows the consequences of each problem and its impact on the business.

Therefore, predicting the failures instead of fixing them is suggested by the researchers. To that end, a business would need to employ a *predictive maintenance* solution driven by AI and machine learning. Furthermore, along the digital transformation journey, the industry needs to adopt wireless sensors and warning systems due to the remote location of the pipelines.

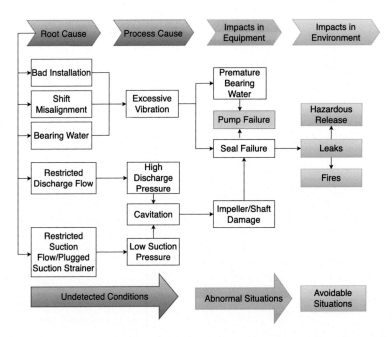

Figure 3.8 A use case of pump failure from root cause to the potential impacts.

3.3.3 Shallow and deep learning methods for pump cavitation detection

Pipelines in the oil and gas industry utilize pumping systems to keep the flow of oil or gas within the pipeline. The pumps mainly suffer from cavitation that reduces the pump's working condition and can lead to failure. Cavitation primarily refers to the forming of bubbles in the pumps that reduces the efficiency of pumps and, consequently, the pipeline system. Therefore, cavitation needs to be recognized early to prevent further losses. Many preventative measures have been taken into account to reduce the effect of cavitation. These detection techniques are based on vibration details and speed variation data. The traditional methods of detecting cavitation primarily focus on declaring the event of cavitation when the total head reduces by some random value (typically 3%) in acknowledgment of a pressure drop at the pump's inlet. However, by the time this acknowledgment has been made, the pump is seriously suffering from cavitation. Hence, the detection of cavitation at an early stage is significantly important for the prevention mechanism.

Advanced technologies are suitable candidates for faster cavitation detection. Two techniques are most efficient in the cavitation detection process. The first technique is utilizing 1 MHz bandwidth acoustic sensor including a dynamic range of 80 dB that captures the in-depth features considering coarse vibration. The second technique is Machine Learning (ML) algorithms that enable signal separation from strong (*i.e.*, pump vibration) to weak (*i.e.*, present in cavitation precursors). Two of the popular ML algorithms to detect cavitation of the pumping system are Bayesian parameter estimation and support vector machine. The ML algorithms mainly target obtaining the important features from the senor-generated data. After that, the ML model is trained with generated data set to detect cavitation of the pump established on a mathematical formulation.

Convolutional Neural Networks (CNN) are also offering effective solutions for cavitation detection. Neural network solutions are extended the problem scope by identification of the intensity of the cavitation in the pumping system [83]. The hydraulic turbines are often damaged by cavitation that happens during turbine operation. The current operating condition can fall under cavitation, and identifying the state of operation is challenging. Hence, detection of cavitation during turbine operation is helpful for the operator to consider a reasonable procedure to reduce or eradicate the impact of the cavitation. In this case, visual identification of cavitation is not possible. Thus acoustic event detection can be a viable solution. The current techniques mainly depend on utilizing acoustic signals to design statistical or handmade features (*e.g.*, kurtosis). The CNN models automatically extract features from the input dataset instead of using handmade features. In addition, it has less dependency on the sensor position.

The occurrence of pump failure for cavitation can be mitigated with continuous monitoring and periodic predictive maintenance of the pipeline system. The modern computing technologies and solution tools (*e.g.*, machine learning, deep neural network) for identifying pump failure need a computing platform for accurate and precise inference or identification operation. For onshore oil or gas fields, cloud computing platforms can support the solution tools to perform accurately. However, in the cloud computing paradigm, one has to send the data to the cloud data center via a network that can cause latency in the inference system. On the contrary, edge or fog computing technologies can be utilized for offshore or even onshore oil and gas fields to serve the computing requirements. The edge servers can be mobile (housed within a vehicle) and transported to remote oil or gas sites to support the computing requirements or even scale up existing computing platforms.

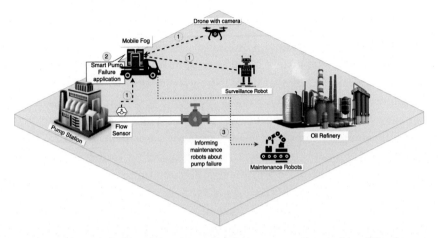

Figure 3.9 Pump failure detection using edge computing and machine learning techniques. Surveillance robots and drones works together with edge/fog computing system for identification of pump failure. Maintenance robots go to the location to fix the problem or change the pump. All the computations can be performed in mobile edge server in the remote site.

Edge-to-Cloud solutions for pump failure detection. The emerging computing technologies (*e.g.*, edge computing, smart IoT sensors) are improving the existing Oil and Gas industry solutions by utilizing machine learning and deep neural network methods. Although, the ML and DNN solutions are computing-intensive and need substantial computing power. Hence, these ML or DNN solutions can be trained on cloud computing platforms (*e.g.*, Amazon SageMaker, Microsoft Azure, IBM) to develop a robust model for inference. Accordingly, as depicted in Fig. 3.9, these pre-trained models are deployed in edge devices (Mobile Fog) that are located in the Oil and Gas industry near the location of the sensors to perform the inference operation in real-time. Sometimes, drones or robots are utilized where the pretrained model is installed for inference. The periodic surveillance and performing inference operations are considered part of predictive maintenance. Fig. 3.9 represents a smart oil field having the issue of pump failure where surveillance robots and drones visit and perform inference operations on received sensor data from the problematic pump. Upon identification of the issue, drones send the signal (step 1) with the location of the malfunctioning pump to the maintenance robots to fix the issue. Then, the mobile fog process (step 2) the received data, and finally, the maintenance robots visit (step 3) the pump failure location and fix or replace the problematic pump. The whole process is less time-consuming and accurate enough to restore the pipeline system's normal procedure, reduce the pipeline's downtime, and increase the production rate.

3.3.3.1 Leakage detection in pipeline

The pipeline is one of the major transport medium in midstream, especially for crude oil. In addition, it is the most convenient and economically viable. However, leakage in the pipelines causes many problems in the midstream transportation, where various leakage detection techniques with the utilization of cutting edge technologies such as machine learning tools, edge or fog computing, IoT devices enable a robust pipeline system. In Table 3.2, various leakage detection techniques are introduced with respect to transportation mediums. In accordance with Table 3.2, we will explore the techniques in the following section:

3.3.3.2 Various computing solutions for pipeline leakage detection

Pipeline leakage has a significant impact on production. Moreover, the surrounding environment gets polluted due to a leakage in the pipeline. Advanced computing technologies can detect pipeline leaks swiftly and accurately. Some of the modern leak detection techniques are explained briefly in the following subsection.

Edge-Cloud-based pipeline leak detection. Most oil and gas companies have a leak detection system in their overall asset integrity management program [84]. Because pipelines are frequently located in rural regions, a leak might go unnoticed for days. Leaks or spills may be expensive, especially if the impacted dirt needs to be removed and treated. There is a significant risk of regulatory fines and lost output. The pipeline Leak Detection Solution may use pipeline pressures, flow rates, and pump status to evaluate flow patterns and identify if a problem exists. The solution presented in Fig. 3.10 can use edge and cloud computing platforms with statistical methods like Sequential Probability Ratio Testing (SPRT), which are widely regarded as among the most trustworthy in the business. This method differentiates between the asset's "regular" and "abnormal" functioning and temporary characteristics. Compared to external systems and visual monitoring, this methodology avoids many of the issues other computational-based methods have and provides a more robust, fast, and hands-off procedure. Rather than modeling the pipeline in mathematical

terms, which may be difficult, the solution can take the approach of ob-
serving the general behavior of the pipeline and characterizing it from a
statistical standpoint to determine the typical and aberrant data patterns.

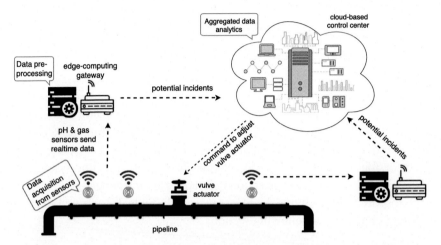

Figure 3.10 An overview of Edge-Cloud-based pipeline leakage detection
system. Sensor generated data sent to edge computing systems for real-time
processing. Whereas potential incidents are sent to cloud services for aggre-
gated long term data analytic that provide control command to valve actuator.

As depicted in Fig. 3.10, the system optimized for single–phase and
incompressible fluids leverages data from current sensors to learn normal
operating parameters and determine when a possible event, such as system
abnormalities or other maintenance concerns, is happening. According to
Fig. 3.10, after data acquisition by sensors, the relevant data are sent to edge
systems for real-time processing. In addition, the system can increase accuracy by tracking pump failures, meter reliability measures, and changes in
pressure and flow using historical and real-time data. Although the software
solution does not remove leaks as part of a leak detection program, early
discovery and response may be able to lessen the impact of leakage.

3.4 Storage facilities for O&G

The total oil products transported from the oil wells to the refinery plant
are not processed immediately. A variety of storage systems are used to store
the oil temporarily before sending them to the refinery plants for further
processing. In addition, the processed and refined petroleum products are

required to store before they are transported and distributed among the end consumers. Typically, storage tanks can be categorized based on the working pressure of the tanks. A type of storage tank that is not pressurized is called Atmospheric Storage Tanks (AST). In ASTs, the pressure inside the tank is atmospheric. ASTs are widely used in the Oil and Gas industry, from the production fields to the refineries. Fig. 3.11 shows the taxonomy of atmospheric storage tanks.

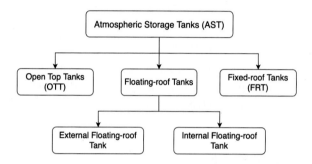

Figure 3.11 A detail taxonomy of atmospheric storage tanks.

3.4.1 Types of atmospheric storage tanks

ASTs are distinguished based on their roof's design and categorized into three main types: Open-Top Tanks (OTT), Floating-Roof Tanks, Fixed-Roof Tanks. In this section, each kind of ASTs is briefly described.

3.4.1.1 Open-top tanks (OTT)

ASTs without any roof are called Open-Top Tanks (OTT). In OTTs, The stored products are open to the atmosphere, and the evaporating losses are high. This characteristic of OTTs limits their use cases. For example, a use case of OTTs is for the collection of contaminated wastewater.

3.4.1.2 Fixed-roof tanks

A type of AST that has a roof that is permanently fixed to the tank shell is called a Fixed-Roof Tank. Fig. 3.12 illustrate a typical Fixed-Roof Tanks. This type of AST is considered the minimum acceptable equipment for storing liquids. Fixed-Roof Tanks can tolerate limited evaporation of products. A pressure-vacuum valve is installed to decrease the evaporating losses due to a slight temperature increase.

Figure 3.12 An example of fixed-roof (cone-roof) and internal-floating-roof tanks. In internal-floating-roof tanks, in addition to the fixed-roof, there exists a roof that floats with the level of the liquid inside the tank.

3.4.1.3 Floating-roof tanks

In accordance with its name, Floating-Roof Tanks have a roof that floats on top of the product (*i.e.*, crude oil, gas). As shown in Fig. 3.13, in Floating-Roof Tanks, the roof moves up and down as the level of the product fluctuates. The main advantages of floating-roof tanks are minimizing evaporation loss and maintaining regulatory compliance for emissions. This makes Floating-Roof Tanks environmentally friendly and ideal for storing flammable products.

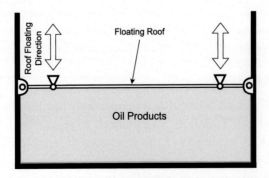

Figure 3.13 An example of floating-roof tanks [85].

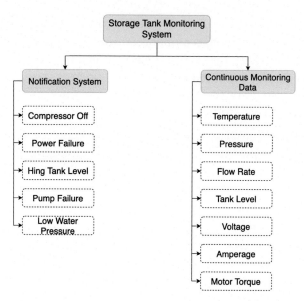

Figure 3.14 Computation in oil and gas storage represented in a taxonomic tree form. The main two subcategories are notification and monitoring systems. Various examples of these two categories are presented in the taxonomy.

3.4.2 Smart O&G storage system

3.4.2.1 Overview

The storage system of the midstream sector primarily includes tanks that can be utilized for processing or refinery works. Oil storage tanks are temporary storage where crude oil from various extraction sites is gathered. Sometimes these tanks are used in production and hold the oil until it is delivered to tankers or into a pipeline. Moreover, storage tanks are also used as a holding area for crude oil before refining. These are also used as storage for refined products after the refining process. Although storage tanks are used for various purposes throughout the industry, the design pattern and usage are typical for all cases. Among various storage systems, Atmospheric Storage Tanks (AST) are the most widely used in Oil and Gas production. ASTs are large above-the-ground storage devices that contain condensed oil and gas liquids. Considering the utilization of computing technologies, we can categorize the midstream storage sector into two areas: AST features and storage tank monitoring systems. A taxonomy of the storage tank monitoring system is presented in Fig. 3.14. As depicted in Fig. 3.14, the storage monitoring system is divided into two categories, namely notification system, and continuous monitoring data. In contrast, AST provides various features to keep the storage system safe and maintain the quality of the products that are presented in Fig. 3.15.

3.4.2.2 Atmospheric storage tank features

An AST should have characteristics that are termed as features of AST in the Oil and Gas industry. AST features typically ensure the quality of the product (QoS) that is stored and enable the safety of the workers and the surrounding environment. To that end, automation processes are adopted in storage systems where state-of-the-art computerized technologies are utilized to enable a robust storage system. Fig. 3.15 depicts the taxonomy of atmospheric storage tank features. This taxonomy explores the automation processes in terms of sensors and actuators.

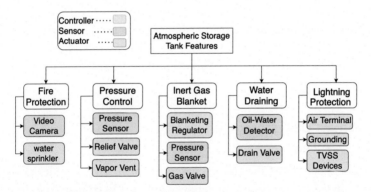

Figure 3.15 Atmospheric storage tank features are presented in terms of automation processes.

3.4.2.3 Fire protection

Oil and Gas storage tanks are often fitted with a ring of nozzles that spray foam into and over the tank's surface in the event of a fire. Although atmospheric storage tank fires are relatively uncommon, they have catastrophic consequences. The fire incident causes a threat to onsite workers and has a significantly adverse impact on the surrounding environment. Hence, ASTs should consider a robust fire protection system, especially automated and computerized, to prevent hazardous situations. Before developing any fire protection system, it is necessary to mention the common fire types that occur in storage tanks. A brief description of 5 typical fire hazards is given below:

Overfill ground fires: Piping or tank leakage can cause overfill ground fires or dike fires. Operator error and equipment malfunction are other causes of such fires. These types of fire incidents are considered the least severe type.

Vent fires: Vent fires are typically associated with Fixed-Roof Tanks such as cone and internal floating roof tanks. A lightning strike ignites fugitive vapors, which may be present at the vent, and causes a fire. Vent fires are a less severe type of fire and can usually be extinguished with a dry chemical fire extinguisher or by reducing the pressure in the tank.

Rim-seal fires: Rim-seal fires happen mostly in external Floating-Roof Tanks and can also occur in internal floating roof tanks or domed roof tanks. As with many tank fires, lightning is the primary cause of ignition; even with floating roof tanks, an induced electrical charge without a direct lightning hit may occur. Because these fires are the most common, there is usually a high rate of successful extinguishment, assuming that there is no collateral damage such as a pontoon failure (explosion) or the floating roof's sinking as due to fire suppression efforts.

Obstructed full liquid surface fires: Obstructed full liquid surface fires can occur in fixed-cone roofs, internal floating roofs, or external floating roof tanks. They tend to be challenging because the roof or pan blocks access to the burning surface. The roof or pan can sink for various reasons, such as an increase in vapor pressure under an internal floating roof, which can cause the pan to tilt. Pontoon failure of external floating roofs is commonly caused by closed drain valves during rains or mechanical seal failure, causing the pan to sink.

Unobstructed full liquid surface fires: Unobstructed full liquid surface fires are relatively easy to extinguish where the tank diameter is relatively small (less than 150 feet) and sufficient resources and trained personnel are available. The most challenging fires will involve larger tanks (greater than 150 feet in diameter) because of the surface area of the fire and the number of resources needed to control and extinguish the fire. Unobstructed full surface fires can occur in Fixed-Roof Tanks without internal roofs, where the frangible weak seam at the roof-shell joint separates due to an explosion or other overpressure event.

Due to technological development and advancement in computing solutions, oil and gas storage fire protection is more robust than before. However, the storage location and the quick spread of fire often create resource scarcity issues. A fire protection system is mainly divided into two phases: fire detection and recovery. The resource constraint issue can cause in both phases due to the synchronization of various activities (*e.g.*, fire detection from video surveillance, sending alert to the affected area, sending notification to the control center) within a short time. Especially, deployment of machine learning or deep learning model needs high configured computation resources that may be difficult to arrange in oil or gas storage due to their remote location. For this reason, edge or fog computing systems with 5G communication technology can provide a state-of-the-art smart solution for fighting the fire in the oil storage tank.

3.4.2.4 Smart fire fighting solution with fog computing and 5G

The fire outbreak incident in the storage tank can cause significant loss and harmful effects or even death to workers. Therefore, a robust and efficient solution is necessary to control the fire swiftly to minimize the loss and mitigate the impact of fire outbreaks in oil storage. Due to the remote location of midstream oil storage facilities, drones, helicopters, and fire fighting trucks are suitable transportation options for a fire fighting scenario that is depicted in Fig. 3.16.

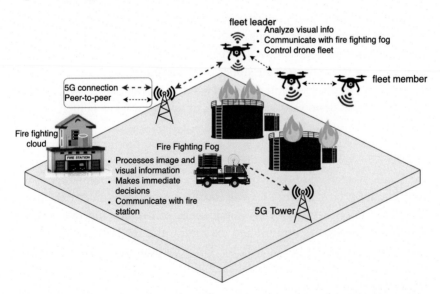

Figure 3.16 Smart fire fighting solution with drone fleet. Fire fighting mobile fog is deployed in the fire location that serves as computing processing unit for fire fighting application.

The smart solution that is presented in Fig. 3.16 utilizes the drone fleet to acquire visual information about the fire hazards in the storage facility. The drone fleet leader sends the estimated hazard information to nearby fire fighting fog and clouds to enable proper safety procedures to mitigate the hazard. In this scenario, fire fighting fog can take real-time decisions to deploy fire distinguishing units in suitable places to reduce the impact of fire hazards.

3.4.2.5 Pressure control system of AST

Atmospheric storage tanks are used throughout the world in the oil and gas industry. These tanks hold liquids until they are moved to the next step in a production or supply chain process. The pressure in these storage tanks is changed due to the temperature and liquid level change. To maintain a safe tank pressure, special valves are utilized to sense small changes in tank pressure. Fig. 3.17 shows an example of AST pressure control system.

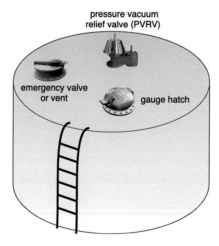

Figure 3.17 Atmospheric storage tank pressure control system components (sensors and actuators).

To maintain safe tank pressure, specific valves that are sensitive to slight variations in tank pressure are used. Several of them are seen in Fig. 3.17. The huge valve in the middle is referred to as a pressure vacuum relief valve in the industry (PVRV). This device is equipped with weighted pallets that open and shut in response to applied pressure. If the pressure within the tank exceeds the predetermined point, the pressure pallet will open, releasing vapors to restore a safe pressure level. If the pressure within the tank falls below a certain level, the vacuum pallet will open, allowing air to enter the tank and raise the pressure back to a safe level. The mechanism is referred to as an emergency valve or vent. In an anomalous pressure scenario, this vent will swiftly alleviate tank pressure. It should be closed under typical circumstances. The little device seen at the tank's front is a gauge hatch for inspection and gauging reasons.

Although the devices mentioned above offer critical pressure control, they have traditionally been unmonitored and lack the feedback loops seen in other pressure control systems. As a result, the number one fear of tank safety engineers and managers, according to a new Emerson Storage Tank Pressure Control Study, is an undiagnosed maintenance problem. Moreover, half (54%) of respondents in a year said they had more than one pressure management concern. Furthermore, these problems were discovered 65 percent after the event.

3.4.2.6 Remote monitoring to control pressure of AST

The remote monitoring option includes a proximity indicator and wireless transmitter. The proximity sensor detects the movement of the vent. "Open" or "Closed" signals are received by the wireless transmitter and can be sent to a control room via a wireless gateway. This option allows quick response to a potential problem.

Safely storing storage tanks is a top priority for tank farms from production to final distribution across the supply chain. In the literature, the benefits of wireless monitoring of tank storage pressure safety valves reflect the importance of pressure management in these tanks and provide feedback to the tank terminal operating staff to help avoid abnormal situations.

3.4.2.7 Inert gas blanket

The tank blanketing, alternatively known as "padding", is the process of filling the empty space of a liquid storage tank with inert gas, most likely Nitrogen. Nitrogen is generally the blanketing gas of choice due to its inert properties, availability, and relatively low cost.

Tank blanketing systems are found on fixed roof tanks as depicted in Fig. 3.18. The system includes a valve that controls the nitrogen coming into the tank. The valve is continuously adjusted to maintain a small constant positive pressure in the vapor space of the tank. Usually, the valve is closed under static conditions, shutting off the flow. If there is leakage in the vessel, the pressure will drop, and to compensate, a low flow of nitrogen into the vessel will occur.

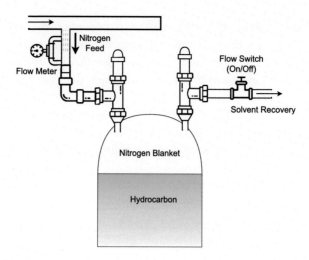

Figure 3.18 The gas blanketing system for oil storage tank.

Similarly, if the temperature drops, there is a decrease in pressure in the vapor space, and nitrogen will flow into the tank. When emptying the tank, significant flow rates of nitrogen can result. In this case, it is essential to maintain positive pressure in the tank. If the tank pump-out rate is higher than the speed of nitrogen flow replacement, negative pressure can cause the tank to suck in and collapse.

3.4.2.8 Water drainage in AST

Storm water accumulated on a concave floating roof of an oil storage tank may affect its floatation, making it necessary to drain the water immediately. A flexible pipe from floating roof down to the bottom of tank, with an outlet above the ground, is used to drain the water.

The necessity of automated drainage system: The drainage of hydrocarbon storage tanks can be a highly random and potentially dangerous operation if it is not automated. This process is traditionally carried out via an operator, who manually opens the drain valves and carries out visual checks to observe the water/oil change. When the operator sees oil in the tank, they close the valve. This method, which requires the presence of an operator in the ATEX zone (potential risks), produces a highly uncertain result.

Problems in drainage operation: The following problems may occur in the operation of a floating roof drainage system:

- The product from the main tank can penetrate the drainage pipe through pinholes or cracks due to aging and drain away with water.
- Product from the tank might run over the floating roof through the roof's seal during overfill, and exit through the water drainage pipe unnoticed.
- Sometimes the flexible pipe is bent or clogged preventing water from the roof to pass through. In this case, water remains on the floating roof and this may disturb its floatation capability.

3.4.2.9 Drainage monitoring applications

Various monitoring applications are developed to overcome the problems related to the drainage system of oil and gas storage facilities. A brief description of some of these applications is as follows:

Detection of oil leaks/spills: An Oil Sheen Detector (*e.g.*, Leakwise ID-223) should be installed in a settling tank (or in a sump or a separator) to collect the water drainage from the roof. The settling tank will settle the liquid flowing in from the roof to allow oil separation and detection by the ID-223 sensor. The normal indication of the Controller should be water on rainy days or air during the dry season. An alarm will be triggered if the ID-223 sensor detects oil or oil on water, indicating that oil is seeping through a fracture in the roof drainage pipe.

Detection of clogs in roof drainage: If air instead of water status is indicated during a rainy day, it means that the flexible roof drainage pipe is bent or clogged, and no water is running through it. This is an early alert that the roof will get too heavy and sink into the tank.

Water treatment costs savings: A Detection System (*e.g.*, ID-223 [86]) can be used to control valves, pumps, and sump gates. Thus, storm water from the tank's roof can be discharged directly to the sea, river, or public drainage system. Only the oily water will be diverted to treatment. This reduces the load on the local treatment system and brings substantial cost savings.

3.4.3 Smart solution—automatic water drainage system

When the crude oil is decanted into the tanks, the water contained in the hydrocarbon is pushed to the bottom of the tank (because it is denser than oil). Before the refining process, it is necessary to drain these tanks to avoid polluting the hydrocarbon. The smart system uses an oil–water interface detector (Minisonic ISD) connected to a control room. When the Minisonic ISD detects a change in the product (mix of oil and water, emulsion that characterizes the end of the presence of water in the tank), the detector sends a signal to the control room using one of the available protocols on the device (4–20 mA, RS 232 or RS 485). This signal orders the valve to be closed, which stops the hydrocarbon from being released (loss of product) and protects against the risk of an accidental oil spill. The water, evacuated via the drain at the bottom, will then be sent to the treatment plant.

3.4.3.1 Hardware equipment for smart solution

The automatic drainage detection system utilizes an interface detector device, namely a minisonic Interface and Sphere Detector (ISD) converter, and a pair of clamp-on probes. The main purpose of Minisonic-ISD is to control the nature and quality of a petroleum product flowing in a pipeline. The relationship between sound velocity and other physical characteristics such as density or concentration permits a very accurate survey about any change in the process.

3.4.3.2 Working principle of the smart solution

The interface detector sends at least three ultrasound waves per second: when more than 3% oil is observed in the water, the speed of the sound in the liquid is affected, allowing the device to detect the interface.

3.4.3.3 Advantages of using smart solution

The automated drainage system has many advantages that create the oil and gas storage facility, safe and smart. The major advantages can be considered as follows:
- Continuous detection, unaffected by noisy environments.
- Quick and easy installation.
- Mounting of probes under load / no shut down required.
- No contact with the fluid / no loss of pressure / minimum maintenance costs.
- Reliable and precise estimation.

3.4.3.4 Lightning protection of AST

Storage tank lightning protection is a highly debated area in the lightning protection field. However, everyone involved in lightning protection agrees that tanks, especially those used in the oil and natural gas field, need lightning protection. An exception is if the tanks themselves are made with the steel of a certain thickness. However, even in those cases, you can never be too careful because of what products tanks are storing. That is why storage tank lightning protection for oil and natural gas takes on a new meaning.

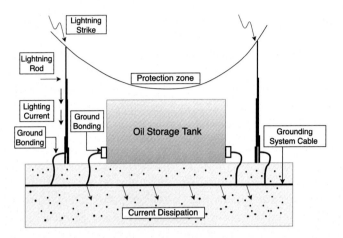

Figure 3.19 Conventional lightning protection system.

Due to conflicting protection criteria, the lightning protection of oil storage tanks has become a controversial subject. One such is the self-protection criteria that rely on the strict fulfillment of the operational conditions, material thickness, and reliable bonding. However, another refers to the necessity of a standalone lightning protection system (LPS) where air terminals are installed (Fig. 3.19) to avoid any interception between the lightning stepped leader and the tank body. Each method has its own advantages and drawbacks. In the past, lightning protection systems (LPS) involved sticking a metal rod into the ground. Nowadays, storage tank lightning protection involves a three-step process:

- Protection of the structure
- Grounding/bonding
- Surge suppression

Structural tank lightning protection employs the basic conventional system with modified air terminals designed to reduce the incidence of direct strikes on the protected structure. In other words, SRATS minimizes the likelihood of direct lightning strikes.

The next step in storage tank lightning protection is grounding and bonding. The purpose of bonding is to bring all the steel equipment to the same electrical potential. Next, grounding is done to ensure that the current escapes to the earth properly. Bonding with grounding is essential to create a safe working environment.

Lightning Master also offers a complete line of subpanel TVSS devices to limit the "sharing" of internally generated transients. These devices are installed on your subpanels. When a transient originating on one circuit travels back to its subdistribution panel, the TVSS device limits it before it can be redistributed onto other circuits within the panel. This approach of installing multiple TVSS devices in series is called "staged protection" and is particularly effective in limiting damage from both externally and internally generated transients.

3.4.3.5 A brief use case scenario of cloud computing in oil refinery

The Industry 4.0 revolution has enabled various solutions for manufacturers, employees, and consumers. For example, the leading oil and gas companies utilize modern computing technologies to maximize resources, improve employee skills, and even identify loopholes before any catastrophic disaster.

The actual benefits of Industry 4.0 are through smart and swift processing of data. The manufacturers have understood the advantage of sensor-generated data that can be turned into valuable feedback insights to improve their products and provide efficient IoT solutions. Hence, one of the concerns is storing the vast amount of data on-premise or in the cloud. Therefore, smart edge and fog computing solutions perform their role as thin computing or storage devices that stay near the data generation source. The cloud computing services for oil and gas, especially data acquisition and analysis, have transformed many European petrochemical refineries. For instance, the Danube refinery in Hungary has 54 primary processing plants that utilize complex processes to convert raw materials into valuable products (*e.g.*, ethylene, propylene, benzene) that are used to make various end products (*e.g.*, resins, fibers, and plastics)

The petroleum refinery plant (among 54 processing plants) processes crude oil using high temperatures to generate gas oil and petroleum coke. Although, the negative side of petroleum processing is the steam eruption that leads to refining the petroleum coke further into a toxic environment. As a result, the whole system needs to be closed to clean the steam eruption-involved equipment. Hence, the steam eruption has negative impacts on workers' health working nearby places.

In petroleum refineries, thousands of sensors are utilized to monitor and control the performance of processing machinery and the quality of the output products. For instance, the Danube refinery generates nearly 100,000 data units every minute that can be analyzed and converted into valuable information. Hence, the refinery utilizes Microsoft's Azure Machine Learning cloud services to perform data analysis. The advantage of using the Microsoft cloud service is the user-friendly interface that allows the IT department of the refinery to perform state-of-the-art analysis with less risk and expenses. For example, using Microsoft's machine learning services, the refinery has already identified several hypotheses to improve the conversion process with less severe steam eruptions. This exploration eventually helped the refinery develop a comprehensive data analysis platform for the in-house employees to perform various analyses on sensor-generated data. Hence, the integration of the cloud services to the operational intelligence system elevates the utilization of more sensor data quickly and with less expense by avoiding the investment of a new system.

3.4.3.6 Smart solutions in oil storage tank maintenance system

Oil storage tanks recognition and estimation of volume using deep learning technique: To determine oil procurement and measure the transportation for oil reservoir strategy, estimating the present amount of global oil reserves and predicting changes in production play a significant role. By measuring the volume of an oil tank, it is possible to estimate the storage capacity of an oil storage base. Detection of the floating head tank and estimation of the reserved/occupied volume of oil present. Followed by reassembling image patches into the full-size image with volume estimations added [87].

Oil storage tank damage detection using support vector machine: The acoustic emission method has a major application in detecting the oil storage tank damage. Therefore, the classification of acoustic emission signals has great significance [88]. A classification method based on support vector machines is proposed for its excellent generalization performance and less training data. The simulation results show that the support vector machine can effectively distinguish different acoustic emission signals and noise signal.

Risk assessment in oil refinery utilizing machine learning method: Oil refineries process [89] hazardous substances under extreme operational conditions to produce valuable products. The required risk assessment is generally time-consuming and involves a multidisciplinary group of experts to identify potential accidents and compute their frequency and severity. In [89], the authors present a machine learning method to mine out useful knowledge and information from available data of past risk assessments. The aim is to automatically classify possible accident scenarios in oil refinery processing units by using SVM (support vector machines). Data from a previous qualitative risk assessment of an ADU (atmospheric distillation unit) of a real oil refinery is used to demonstrate the applicability of the SVM-based approach. The test classification was made with an F1 score of 89.95%. In this way, the results showed that the proposed method is promising for efficiently performing an automated risk assessment of oil refineries.

3.5 Safety in midstream sector

3.5.1 Overview

The safety concern of midstream mainly has two types of impacts where various computations are utilized to reduce the outcomes. These two types of safety are typically workers' personal safety and environmental safety. The construction and operation of pipelines, rail tracks, compressor stations, storage facilities, and terminals can have significant ongoing impacts on the environment and human health. The midstream sector of O&G industry typically transfers extracted materials from oil or gas rig sites to refineries or processing plants. The Midstream sector is one of the most complex networks of supply chains in any industry where miles of pipeline, the platoon of trucks, tanker ships, rail cars, barges, acres of tank farms, equipment, and many other fault intolerant operations are synchronously functioning to make the oil and gas transfer safe. This sprawling and complicated system exposes midstream workers to numerous safety hazards at every stage. Therefore, the transportation system and processing facilities need to be safer for workers and the surrounding environment.

3.5.2 Workers safety in midstream

3.5.2.1 The impact of toxic gas on workers health

In midstream, processing operations are used to separate the byproducts and wastes from raw natural gas to produce "pipeline quality" or "dry" natural gas for consumption [90]. Natural gas processing plants have the potential to release significant quantities of air pollutants that can be harmful to on-site workers. Table 3.4 shows various toxic gases that can be released due to leakage or malfunction of different processes in midstream. These emissions result from the combustion of natural gas, fossil fuels in compression engines, the fugitive release of volatile organic compounds (VOCs), and Sulfur Dioxide. In addition, hazardous (*i.e.*, toxic) air pollutants (HAPs) from separators, dehydrators, and sweetening units used to extract byproducts, and wastes from the natural gas stream create unhealthy surroundings for workers.

Table 3.4 Various toxic gases can be released due to leakage or malfunction of different processes of midstream. These toxic gases and how they can impact the human health are depicted in a tabular format.

Toxic Gas	Symbol	Safety and Environmental Side Effects
Volatile organic compounds including benzene, formaldehyde	VOCs	VOCs can cause cancer. VOCs react with NOx to form ozone, a respiratory irritant, and greenhouse gas.
Particulate matter	PM	Affects the heart and lungs
Methane	CH_4	Main component of natural gas. Much more powerful than CO_2 as a greenhouse gas.
Carbon dioxide	CO_2	Major greenhouse gas
Nitrogen oxides	NO_x	React with VOCs to create ozone
Hydrogen sulfide	H_2S	Can cause illness and death at certain concentrations

3.5.2.2 Sources of toxic gas emission

There are various emission sources for toxic gases, including oil and gas refineries, storage, and processing plants. The quantity of emissions can be different depending on the operations of the oil and gas industry and the geological locations. While the emission is regular in some cases, it is intermittent in others. This section of the book considers the cases for emissions related to the midstream sector. Typically, the extracted oil is transferred to a refinery where gas and condensates are separated and processed. This processing of crude oil has many subprocessing that includes dehydration, heating treatment operation, and finally compression. All these subprocessing operations are a potential source of toxic gas emission that is depicted in Fig. 3.20.

3.5.2.3 Edge-to-cloud solutions for toxic gas detection

Automation techniques are implemented in midstream processing plants to protect the workers from exposure to various toxic gases. Sometimes, these processing plants are located in remote areas where network connectivity is unstable. Due to this stochastic connectivity Cloud computing platform is not a suitable platform for performing fault intolerant and real-time automation processes. A potential alternative is edge computing that can be functional in remote areas in a standalone fashion without connectivity to a cloud data center. As depicted in Fig. 3.21, the edge devices/nodes can

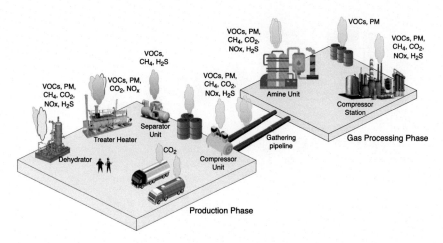

Figure 3.20 Various sources of toxic gas emission are depicted with subprocessing unit in refineries and pipelines. This diagram shows typical oil processing steps with potential toxic gas emissions that are harmful for workers and people living nearby. Although, the equipment and processes can be different for different operating midstream companies.

download (step 1) toxic gas detection DNN-based models [91] when the network is stable and do the inference afterwords.

The maintenance workers carry toxic gas detection sensors with their protective safety suits while working in a hazardous area. In a toxic gas emission incident, the sensor detects the toxic gas element in the air and sends sensor data to a nearby processing edge server (step 2). The edge server already has the toxic gas detection and intensity level machine learning (ML) model. With the sensor data as input, the model can perform toxic gas detection and intensity level measurements (step 3). According to Fig. 3.21, if the intensity level goes beyond the harmful level, then the edge server notifies (step 4) the rescue team as well as the victim workers for safety procedures.

3.5.2.4 Motor vehicle accidents

A significant portion of the oil and gas industry's midstream sector transfers the extracted oil and gas from the production well to downstream distribution centers. In this transfer, motor vehicles are widely used where accidents can occur for various reasons (*e.g.*, bad weather, dumps on the roads, malfunction in vehicles). Due to motor vehicle accidents in midstream transportation, many human workers fall victim to injuries; even

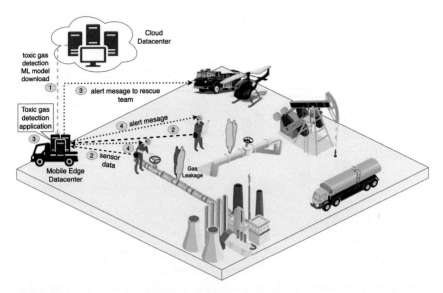

Figure 3.21 In an oil extraction site maintenance workers carry toxic gas detection device that can detect the toxic gas emission. The sensor data processing is performed on mobile edge datacenter that has the machine learning model to detect the toxic gas and it's intensity level. For harmful level of toxic gas emission, the edge device sends alert message to worker and if necessary sends messages to rescue team.

death may occur. Reducing human activities (*i.e.*, autonomous vehicles) in transportation and early detection of various flaws in the transporting mediums (*e.g.*, trucks, barges, rail cars) can alleviate motor vehicle accidents. The autonomous vehicular systems and malfunction detection in transporting vehicles utilize state-of-the-art computing solutions and processing platforms to make the midstream sector safe for human workers. In addition, the emerging field of deep learning and IoT systems are developing various robust solutions with high accuracy and less power consumption ability.

The autonomous vehicle utilizes various machine learning and deep learning models for image classification, object detection, semantic segmentation, and character recognition systems that ensure workers' safety in the midstream sector. In disasters or accidents, avoiding affected roads need precise and accurate computation to find the suitable path to the destination. Additionally, various vehicular applications (*e.g.*, forward collision avoidance, pedestrian detection, adaptive braking systems, platoon coordinating system) reduce the motor vehicle accidents in the midstream sector.

3.5.2.5 Fire hazards

The Oil and Gas industry is one of the riskiest industries regarding the health and safety of employees. Workers in the Oil and Gas industries face the risk of fire and explosion due to the ignition of flammable vapors or gases. Wells, trucks, production equipment, or surface equipment such as tanks and shale shakers can release flammable gases. Ignition sources can include static, electrical energy sources, open flames, lightning, cigarettes, cutting and welding tools, hot surfaces, and frictional heat. With the flammable chemicals, gases, and materials used in the Oil and Gas industry, employers are paying great attention to fire protection in the workplace. Controlling fire hazards is a crucial part of ensuring worker safety in the midstream sector of the Oil and Gas industry. Fire hazard incidents can be swift enough to cause an explosion, and the spread of fire can be quite fast, which can cause serious health problems for workers in the disastrous area. Hence, a real-time fire detection system must synchronously incorporate various recovery activities. For example, automated water sprinklers and fire extinguishers should be in hazard-prone areas. In addition to these systems, many different computing activities should collaborate swiftly and accurately to mitigate the fire hazard of a larger fire and explosion occurrence. For instance, drones or surveillance robots should periodically visit the potential hazard-prone areas and capture pictures or videos to identify any anomaly that can cause a fire. In the case of remote extraction sites, edge or fog computing servers can be deployed with an anomaly or fire detection machine learning model that receives surveilled images/videos from drones and surveillance robots and perform inference to detect fire. Upon identification of fire, the fog system informs the recovery team and fire extinguisher robots to mitigate the incidents.

3.5.3 Environmental impacts of midstream

3.5.3.1 Methane emission

Methane emissions are not only wasting resources but also causing the greenhouse effect. EPA (U.S. Environmental Protection Agency) has reported that natural gas and petroleum systems are the largest sources of methane emission in the U.S. As a result, accurate localization and fast detection of a methane leakage event are of crucial importance.

A Wireless Sensor Network (WSN) technology proposed by [92] is used to monitor any methane emissions in the gas well pad continuously. The proposed WSN is composed of methane, wind, and temperature sensors, a data processing board (Raspberry PI), GPS unit, a cellular modem, and a solar panel for power supply. When the WSN is installed, it begins data acquisition automatically. Then the collected data is streamed to the cloud. Shortage of power supply and limited bandwidth are the main constraints for deploying WSN in remote areas. Power management and Edge computing are two solutions employed to enable WSN in remote areas. Finally, the back-propagation method is presented to locate the source of the methane leakage. According to this method, it is assumed that methane leakage propagates and spreads along the wind direction line. Suppose two sensors detect a methane peak (not necessarily at the same time). In that case, the back-propagation method locates the leakage source at the intersection of wind direction lines detected at each sensor location.

3.5.3.2 Carbon emission

The Oil and Gas industry has considerable prospects to be a game-changer in meeting climate-change goals. The industry operations account for 9% of all human-caused greenhouse gas (GHG) emissions. It also generates the fuels that account for another 33% of world emissions. CO_2 emissions from energy and material production can come from various sources and fuel types, including coal, oil, gas, cement manufacturing, and gas flaring. Moreover, the proportion of different fuel sources to CO_2 emissions has altered geographically and temporally as global and national energy systems have transformed throughout centuries and decades.

Carbon emissions from flaring: During the oil extraction process, flares are used to burn off natural gas in a controlled manner. Otherwise, natural gas can burn out of control, posing a serious threat. Natural gas is usually trapped, but it is flared when that isn't possible to utilize safely for an appropriate purpose. Flaring decreases the risk of gas ignition in facilities and helps to eliminate items that are no longer appropriate for use [93]. The volume of gas flared off is usually measured in billion cubic meters (bcm). Like burning natural gas as a fuel, natural gas flaring releases carbon dioxide into the environment. One of the significant problems with flaring is that it has no actual purpose or function other than to keep people and equipment safe. Except for a few nations, flaring rarely contributes to a large amount of a country's greenhouse gas emissions (mostly in Africa). Flaring accounts for less than 1% of total CO_2 emissions across the world.

Flaring emits vast amounts of greenhouse gases into the environment while producing no work (useful energy). On the other hand, flaring the gas has a less significant global warming effect than letting the natural gas escape into the atmosphere. The major component of natural gas, methane, has a more considerable global warming potential than carbon dioxide. As a result, by burning methane instead of producing carbon dioxide, the amount of carbon released into the atmosphere is reduced. However, this does not imply that flaring is advantageous because retaining the carbon underground would have a much less impact.

Many oil producers are investigating strategies to lessen the need for flaring as global warming becomes a more generally acknowledged concern. As a result, flare-ups have diminished over the world. New technologies are being developed that will allow natural gas to be used in ways it has never been before. Natural gas, for example, may now be injected back into oil wells to raise pressure and allow oil to be produced continuously [94]. Furthermore, as natural gas has become a product that firms seek to harness and sell, the quantity of flaring has decreased even further [95]. Cryptocurrency mining, which has lately drawn extensive attention to the oil and gas sector, is an interesting example of flaring reduction. "Crusoe Energy," a leading natural gas flaring solution provider [96], utilizes gas that would otherwise be squandered to power computers that perform intensive processing tasks such as Bitcoin mining. According to Crusoe, this method "reduces CO2-equivalent emissions by 63% compared to continuing flaring." According to Bloomberg [97], ExxonMobil and Crusoe Energy have been working on a trial project in Dakota's Bakken plains to sell surplus gas to third-party Bitcoin mining companies. In a report, Consumer News and Business Channel (CNBC) states that ConocoPhillips was testing a similar strategy in the same region. Flaring, the procedure of burning off surplus gas created during oil extraction, is under rising pressure from oil and gas companies. According to the World Bank, almost 142 billion cubic meters of gas will be flared in 2020, resulting in roughly 400 million tonnes of CO2 equivalent emissions. As a result, the bank has urged oil companies to stop using the method by 2030. Therefore, using excessive gas for Cryptocurrency mining can reduce the impact of carbon emissions from oil extraction sites of the O&G industry.

3.5.3.3 Oil spill

The release of liquid petroleum into the marine environment is a form of water pollution. However, oil leakage is not limited to the marine environment and can also occur in transport pipelines on land. The ecosystem response to the 1989 oil spill in Alaska was investigated by [98]. The study indicates that the unexpected persistence of toxic subsurface oil has a long-term effect on wildlife. A novel approach consisting of UAVs and computer vision is proposed by [99] to automatically and efficiently detect oil spills. In this work, With a well-trained DCNN model, oil spills can be detected automatically and precisely. The model produces a probability value after an aerial image is captured. The model indicates no oil spill only if the probability value is less than a preset threshold. The UAV continues to patrol. There may be an oil spill once the value is higher than the preset threshold. When using a well-trained DCNN for oil spill detection, the DCNN shows high recognition accuracy. However, false alarms are still possible. For example, when the central part of a polluted river and oil share the same characteristics, the network may detect an oil spill improperly, leading to a false alarm. Nevertheless, it is possible to employ the Otsu algorithm to distinguish between oil spills and other interferences. In the previous example, the change in the brightness of the polluted water is a gradual process, while the relative brightness of the nonoil spill area compared to the oil spill area produces a dramatic change. The Otsu algorithm can tell the difference between the two. When DCNN detects an oil spill area, an interference removal program will run automatically to eliminate the false alarm.

3.6 Security aspects of midstream sector

Exploring the processes involved in midstream is necessary to understand the security issues of midstream. In the Midstream sector, the extracted oil with impurities passes through local pipelines and is pumped into primary reservoirs-batteries where oil is separated from gas and water [100]. After that, the crude oil is first stored in storage tanks, and then it travels through oil trunk pipelines to oil refineries and other storage tanks, tanker vessels, or tank wagons for transportation. Pumping stations are installed to pump oil through pipelines at regular intervals along the entire route length. Pumps are used to initiate and maintain pressure, overcome friction, and account for the difference in altitude along the route length and other factors.

The Industrial Internet of Things (IIoT) and the influx of increased connected devices online and greater data demand have caused midstream applications to increase the adoption of digital systems, predictive maintenance service models, and more to address global energy demands. As a result, cyber security threats continue to rise against such midstream applications as Power Equipment Centers (PECs), compressor stations, LNG terminals, pipelines, and pump stations. Factor in the reality that those cyber security threats add to the already typical safety challenges and risks facing midstream oil & gas—such as explosions, fires, spills, and more. It's no surprise that midstream operations are taking proper precautions to minimize downtime, secure continuous, efficient operations, prevent environmental damage in the wake of a possible catastrophe, and protect the integrity of their critical infrastructure.

According to the midstream sector taxonomy presented in Fig. 3.1, the security aspects of in midstream sector are divided into two categories, namely cyber and physical. Cyber security threats mainly refer to the issues where the physical presence of an intruder is not found. The cyber-attacks mainly refer to digital hacking or digital theft. On the contrary, physical security threats involve the presence of an intruder inside or near the oil and gas extraction premises to damage any site asset.

3.6.1 Cyber-attacks on O&G midstream

The emergence of smart oil fields and the Industrial Internet of Things (IIoT) have transformed the O&G industry significantly while making it more prone to cyber-attacks. Considering a report of PwC's Global State of Information Security Survey of 2016, 42% of O&G companies fall victim to phishing attacks. In addition to the history of attacks in the oil and gas industry, these surveys clearly state that oil & gas is a highly targeted industry. A brief history of cyberattacks against O&G industry is provided in the following section.

3.6.1.1 A brief recent history of cyber-attacks against O&G industry

- In 2010 STUXNET [101] virus was utilized to attack industrial control systems all over the world that includes computers in refineries, pipelines, and power sector.
- In 2012 a huge cyberattack was occurred in Saudi Arabia that result in damaging around thirty thousand computing devices. The main focus of the attack was to break down the resource transportation from Saudi to international market.
- In 2012 internal firewall and security system breach occurred in the energy sector; specifically, the remote administration and monitoring system was the target. The hacker was able to extract the project files related to the SCADA project. In this case, the main focus was to acquire the source code of the controlling software to get hold of the system.

- The Ugly Gorilla attack was occurred in 2012 that hacked around 24 U.S. natural gas utilities. The attacker was able to steal sensitive information from oil and gas pipeline companies.
- A different kind of malware that can record audio, screenshots, and user activities was used to attack the Middle Eastern Company.
- In 2014 an attack was triggered via email targeting approximately 300 different firms within the Norwegian Oil & Gas industry. The attachment of the email installed to the system when user opened it, and explored the loop holes of the security system.
- Smart Automated Tank Gauges (ATG) could be remotely accessible by the hackers, and shutdown the whole system within few minutes. This type of attack was made in 2015 to the U.S. oil and gas industry.

3.6.1.2 Cyber-attacks in various sectors of O&G

Cyber-attacks are taking place in every sector of O&G industry, where midstream is the most vulnerable sector to the attackers. From the timeline history of cyber-attacks mentioned earlier, categorizing the cyber-attacks across the Oil and Gas landscape provides a better overview of the attacker's primary target. As such, we categorize the cyber-attacks reported in the past according to their target sector in a tabular form in Table 3.5.

Table 3.5 Reported cyber-attacks according to their target sector.

Cyber-Attacks	Upstream	Midstream	Downstream
PDVSA	yes	yes	yes
Baku-Tbilisi-Ceyhan		yes	
Bayamon		yes	
STUXNET		yes	yes
Saudi Aramco	yes	yes	yes
Telvent	yes	yes	yes
RasGas		yes	
Flame	yes	yes	yes
Norwegian Attacks	yes	yes	yes
Statoil	yes		
ATG			yes

3.6.1.3 Cyber-attack target area—SCADA

Cyber-attackers mainly focus on targeting Supervisory Control and Data Acquisition (SCADA) systems that primarily control complex oil and gas processes in various steps. For example, the SCADA system manages the operational parameters (*e.g.*, pipeline pressure, transfer speed, temperature) related to transferring hydrocarbon from one location to another.

A successful attack that breaches into internal networks of midstream control system can remotely control relief valves and compressors. Moreover, attackers can override automatic shutdown systems that protects against over-pressurization of pipelines. Furthermore, the situation can be worse enough to kill onsite workers operating near the pipeline stations.

3.6.1.4 Prevention from cyber-attacks

Many midstream companies are improving their cyber security system to avoid attacks. IT experts utilized most state-of-the-art firewall systems that mainly focus on proactive defense strategies, including detection, monitoring, and assessing the vulnerable loopbacks.

Considering the security threats, cloud computing provides solutions for oil and gas companies to develop frameworks for controlling systems distributed across various servers, including independent security strategies. This distributed security system effectively avoids security breaches to gain control of the whole operating framework. Therefore, the probability of massive failure of infrastructures can be mitigated using the distributed security system.

3.6.2 Physical threats of O&G

The physical threats of the oil and gas industry mainly refer to the threats to the infrastructures of midstream operations. The final results of physical threats are physical attacks on the midstream infrastructure to destroy some properties of the midstream sectors. Hence, various terrorist organizations, especially international terrorists, have a history of conducting physical attacks on the oil and gas industry. Some listing of the physical attacks is given below:

- Several groups in Columbia attacked the national oil pipeline so frequently that it has become fairly nicknamed "the Flute" [102].
- One group in India in 2006 attacked the oil pipeline of Assam, a province of India. The pipeline is the source of around fifteen percent of India's onshore oil production.
- One of the Mexican groups performed six parallel attacks on Mexico's oil and gas industry that caused a severe supply shortage. Eventually, the temporary shutdown of the facility [103] impact the economy significantly.

3.7 Summary

The midstream sector of oil and gas is the bridge between upstream and downstream sectors. The midstream mainly focuses on transportation and temporary crude oil storage and other products. Hence, the midstream sector can improve the transportation of crude oil by utilizing advanced motor vehicle technologies and pipeline management systems. However, another sector like petroleum storage can enhance its operational procedure by using machine learning and other software tools. Various advanced technologies of deep learning, artificial intelligence, smart sensors, actuators, and high-performance computing platforms make crude oil transportation efficient and profitable. The involvement of advanced computing is driving the midstream sector to become more autonomous, cost-efficient, and environment-friendly.

CHAPTER 4

Smart downstream sector of O&G industry
Smartness in downstream sector

4.1 Introduction and overview

The operating procedure of the downstream oil and gas industry mainly refers to the last step of the supply chain process where crude oil and natural gas are refined and processed into many end products and transported to the end consumers. Hence, the final sector of the oil and gas industry is known as 'downstream,' which includes everything involved in turning crude oil and natural gas into thousands of finished products utilized in various daily life activities. Fuels like gasoline, diesel, kerosene, jet fuels, heating oils and asphalt are some major products of the downstream. Moreover, long-chain hydrocarbons that exist in both oil and gas are processed to create less utilized products like synthetic rubbers, fertilizers, preservatives, containers, and plastics.

In addition, oil and natural gas products are even used to make artificial limbs, hearing aids, and flame-retardant clothing to protect firefighters. In fact, these are just some examples of products derived from the crude oil or gas. There are many more petroleum products (*e.g.*, paints, dyes, and fibers) that their quality mainly depends on efficient downstream operations.

The entire downstream operation along with upstream and midstream presented in Fig. 4.1 as a flow diagram. Conversion of crude oil and gas into the end products (*i.e.*, oil refining and gas processing) is generally classified as the downstream operations. Oil refining operations (*e.g.*, refining crude oil into gasoline or diesel) mainly occur in the downstream sector. However, gas processing is sometimes performed in midstream. Hence, the midstream and downstream of the oil and gas industry are closely related to each other as depicted in Fig. 4.1. For this reason, large oil companies are described as "integrated" because they combine midstream oil and gas processing with downstream operations. More specifically, the refining process occurs either in midstream or downstream, and the distribution of oil and gas occurs in the downstream phase.

IoT for Smart Operations in the Oil and Gas Industry
https://doi.org/10.1016/B978-0-32-391151-1.00013-7

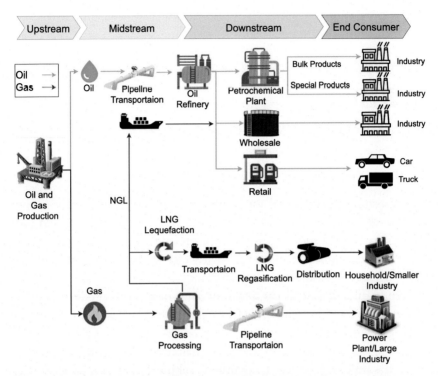

Figure 4.1 Flow diagram of petroleum products, focusing on downstream operations. The green lines are considered as oil whereas blue lines are gas. The flow of oil and gas starts from left of the figure, and finishes with end consumers.

The exposure to end–users from O&G industry mainly occurs via downstream companies. Hence, this sector can be considered as a bridge between end consumers and O&G industry. Once crude oil is located and extracted in upstream, it is shipped and transported within the midstream process. Next, the oil is refined, marketed, distributed, and sold, which are classified as the downstream sector. Fig. 4.1 depicts the whole value chain of oil and gas products from upstream to downstream with more emphasis on downstream.

4.1.1 Downstream taxonomy

The overall downstream sector is depicted in Fig. 4.2. The downstream sector is mainly categorized into three subsectors: refining and processing, distribution, and safety. The majority of the downstream sector mainly

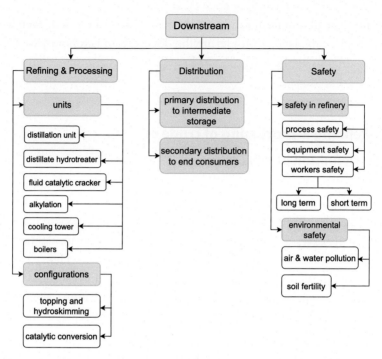

Figure 4.2 The taxonomy of downstream sector. The major sub categories are refining, distribution, and safety.

refers to refining and processing crude oil and natural gas into end products. The refining process includes units and configurations through which refining plants transform crude oil into petroleum products (*e.g.*, diesel, kerosene, jet fuel). Next, in the distribution subsector of the downstream, petroleum products are distributed to intermediate storage and end consumers. Safety requirements are essential practices in downstream. Safety in refinery and environmental safety are two major categories of the downstream safety. The primary safety operations in refinery safety are process safety, equipment safety, and workers safety. Finally, the environmental safety is of crucial importance in downstream. Oil and gas refineries are major sources of hazardous air and water pollutant that directly impacts the surrounding environment. That is why we need to be careful about air and water conditions near oil and gas refineries. Furthermore, while producing refined end products, it is possible to generate toxic chemical wastage as a by-product. Hence, disposal of the chemical wastage of infertile land or near any agricultural farm can cause harmful effects to the soil, even losing the fertility of the land. Considering the downstream taxonomy, we discuss it more thoroughly by traversing its branch following a top-down manner in the next sections.

4.2 Refining & processing of crude oil

4.2.1 Definition of O&G refinery

Oil refining is an indispensable process for transforming crude oil into end products such as fuels, lubricants, and kerosene. Hence, an oil refinery is a petrochemical plant that refines crude oil into various usable petroleum products. It serves as the second phase in the crude oil production process following the actual crude oil extraction in upstream. Hence, refinery services are considered as a major subcategory of the downstream.

The first step in the refining process is the distillation. This process includes heating crude oil at extreme temperatures to separate the different hydrocarbons. Oil refineries are typically large, with extensive piping running throughout, carrying streams of fluid between large chemical processing units, such as distillation columns. Various configurations of the refinery units are utilized to produce different end products. Two of the widely used configurations are discussed in the following section.

4.2.2 Different configurations of petroleum refineries

In oil refineries, different configurations are utilized to process particular raw materials into the desired utility product. Using a survey operation on the local market to demand petroleum products and analyze the accessible raw materials, engineers and executive officers set the most economical viable configuration. Since about half the fractional distillation product is residual fuel oil, the local market for it is of utmost interest. Parts of Africa, South America, and Southeast Asia have a great demand for heavy fuel oil. Accordingly, refineries of simple configuration may be sufficient to meet the demand. Nevertheless, in the United States, Canada, and Europe, large gasoline quantities are in need. Moreover, the environmental safety regulations and availability of natural gas put constraints on the fuel market. Hence, more complex configurations are imposed for maintaining the rules and quality of the product.

4.2.2.1 Topping and hydroskimming refineries

The most straightforward refinery configuration called topping refinery (depicted in Fig. 4.3) is designed to provide feedstocks (*e.g.*, naphtha, natural gas) for petrochemical manufacture or industrial fuel production in remote oil-production areas. This refinery system consists of tankage, a distillation unit, recovery facilities for gases and light hydrocarbons, and critical utility systems (steam, power, and water-treatment plants).

Refineries using topping configuration produce large quantities of unfinished oils and highly dependent on local markets. The addition of hydrotreating and reforming units to this basic configuration results in a more manageable hydro skimming refinery that can also produce desulfurized distillate fuels and high-octane gasoline. Hydro skimming configuration can also produce significant residual fuel as end products. Additionally, it may fall into an economic disaster if demand for sulfur-rich fuel oils decreases.

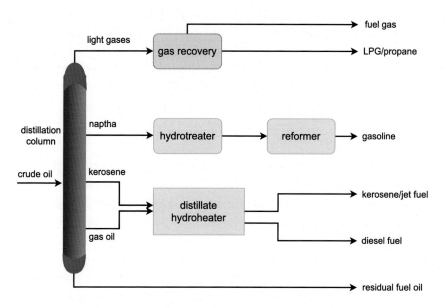

Figure 4.3 Flow diagram of a refinery in topping configuration.

4.2.2.2 Catalytic conversion refineries

Catalytic conversion refinery or conversion refinery is the most common refinery design. The essential building components found in both the topping and hydro skimming setups are included in a conversion configuration.

This structure also includes gas oil conversion facilities such as catalytic cracking and hydrocracking units, olefin conversion plants such as alkylation or polymerization units, and, in many cases, coking units to reduce or eliminate residual fuel output drastically. Modern conversion refineries may generate two-thirds of their output as gasoline, with the remainder split among high-quality jet fuel, liquefied petroleum gas (LPG), diesel fuel, and a trace of petroleum coke. Many of these refineries also employ solvent extraction procedures to make lubricants and petrochemical units, which may be used to recover high-purity propylene, benzene, toluene, and xylenes for further processing into polymers.

4.3 Petroleum refining processes

Refining processes break crude oil down into various components that are then selectively reconfigured into new products. Petroleum refineries mainly transform crude oil into fuels used for transportation, heating, and generating electricity. Besides, refineries produce feedstocks for making chemicals. Petroleum refineries are complicated and expensive industrial facilities maintaining various processes and steps for producing petroleum products. All refineries have three basic steps that are as follows:

4.3.1 Separation

In separation process, the crude oil is first heated to produce a mixture of gases with different boiling points. Then, distillation units are utilized to extract distinct products from the mixture. Specifically, crude oil passes through a pipe to a hot furnace for heating. Then, the outcome of the furnace is released into distillation units. Typically, atmospheric distillation units are found in all refineries, whereas high facility refineries can have vacuum distillation units. The distillation units mainly separate the outcome of the furnace into petroleum components, namely fractions. Depending on the boiling points, fractions are separated into different trays. More specifically, weighty fractions remain at the bottom while light fractions stay on the top.

Distillation towers can have three different chambers. These three chambers are used to capture lightest (*e.g.*, gasoline, liquefied natural gas), medium (*e.g.*, kerosene), and heavy (*e.g.*, gas oils) weight fractions from top to bottom, respectively.

4.3.2 Conversion

The medium-weight and heavy-weight fractions captured in the distillation tower have the potential to be separated into lighter products. The conversion process converts heavy-weight fractions into light products like gas, gasoline, and diesel. In general, the conversion method is termed as "cracking", which utilizes heat, pressure, and catalysts. In some cases, hydrogen is utilized to crack heavy-weight fractions into light products. A typical cracking unit uses reactors, furnaces, heat exchangers, and other vessels. Other than cracking, various refinery processes (*e.g.*, alkylation, reforming using heat, pressure, and catalysts) are performed that reorganize the molecules to increase the product's value.

4.3.3 Treating

Treating is the process of removing or decreasing corrosive and pollutant molecules (sulfur) from end products. This is the final step in improving the quality of final products. For instance, technicians in the refinery cautiously fuse various streams of the processing units. In addition, the level of octane, vapor pressure ratings, and other particular analysis influence the gasoline blend.

4.4 Smartness in refinery units

4.4.1 Distillation unit

The oil refining processes generally start with distillation units. According to the taxonomy shown in Fig. 4.2, this is the first unit of the "units" subcategory. The main objective of distillation is to retrieve the light substances from the crude oil. The process begins with a *fractional distillation column* which separates various components of the crude oil by considering the size, weight, and boiling point of individual substances.

As presented in Fig. 4.4, crude oil from refinery storage is sent to the Furnace to increase the temperature. In this stage, high-pressure steam is utilized to heat the raw crude oil and evaporate it, which is mentioned as the Furnace stage. Then, different trays of the fractional distillation column capture the vapors as depicted in Fig. 4.4 with the distillation unit stage. Since substances in the mixture have distinct boiling points, vapors are captured in different trays. Specifically, the lowest boiling point substance is condensed at the top of the column, whereas the highest boiling point substance is condensed at the bottom of the column. Finally, the trays

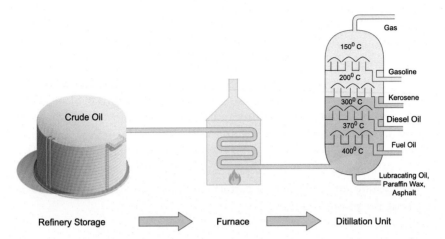

Figure 4.4 A typical distillation column with processing the crude oil into various end products.

collect their designated liquid fractions to pass to cooling condensers or directly to storage tanks, where they undergo further chemical treatment to improve the quality.

4.4.2 Smart distillation

The energy efficiency of processes and operations in a refinery is of crucial importance. According to a US Department of Energy report, 794 TBTU (Trillions British Thermal Units), which equals 26% of the consumed energy, could be saved yearly if the current R&D-based technologies were effectively implemented in US refineries [104].

The distillation process is one of the energy-intensive units in the petroleum refinery system. Therefore, a smart distillation process aims to reduce energy consumption while efficiently processing crude oil utilizing state-of-the-art technologies (*e.g.*, artificial neural networks, cloud computing, IoT). Durrani et al. utilize Artificial Intelligence to design an energy-efficient operation in crude distillation units (CDU). A brief description of the work is presented in the following subsection.

4.4.2.1 Artificial intelligence method for smart distillation

Petroleum refineries utilize many energy-consuming processes, among which crude distillation unit (CDU) is one of the most widely used and energy-intensive processes. The uncertainty worsens the situation by incurring a huge amount of energy wastage. To reduce the energy wastage, artificial intelligence techniques have been studied by researchers to predict the composition of crude oil to utilize a suitable distillation process. Researchers could develop a multi-output artificial neural network (ANN) model that can predict the composition of crude oil to synchronize with the uncertainty behavior. The suggested method is an extension of earlier method of cut-point optimization considering hybrid Taguchi [105] and genetic algorithm. To train the model, a synthetic dataset was generated with hundreds of variations of crude compositions and their optimized cut point temperatures. Empirical results and a simulated CDU flow-sheet were used to validate the proposed model. The results show that the method is faster than baselines with less computing expenses. Therefore, this smart method for efficient distillation in smart refineries is essential to reduce the energy consumption of distillation processes.

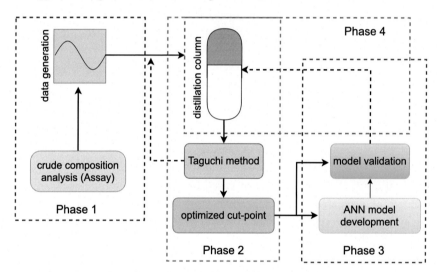

Figure 4.5 Crude composition prediction model development process diagram for smart distillation. The whole process is divided into four steps that are presented into four phases.

Fig. 4.5 represents a schematic diagram of developing the ANN model with the integration of a hybrid Taguchi framework. The whole process is divided into four phases. The very first phase (phase 1) starts with data generation by inserting variations into the crud composition of a Zamzama crude/assay [106]. The second phase estimates the optimized cut-point temperature using the hybrid (*i.e.*, Taguchi and GA) optimization framework. Then the third phase develops the ANN model utilizing the datasets, including the compositions and their related cut-points. Finally, according to Fig. 4.5 phase four, the ANN model is then used to predict cut-point temperatures for any varied composition.

4.4.2.2 Computing solutions for smart distillation

Recent computing technologies (*e.g.*, cloud computing) can improve the efficiency of the proposed artificial intelligence model. For instance, machine learning models can utilize cloud computing facilities and big data storage in the training stage. The services offered by cloud providers can accelerate the end-to-end development of the system. From the security perspective, federated learning [107] with fog federation [108] can utilize data from different companies to train a machine learning model without sharing the data. Hence, the machine learning model becomes more reliable and robust with various companies' data sources and efficient utilization of latency-aware computing devices (*e.g.*, edge, fog, and IoT devices).

4.4.3 Cooling towers

The cooling towers utilize a water cooling system to remove heat from refineries. The mechanism of cooling tower is presented in Fig. 4.6. A typical refinery employs two types of cooling towers: direct (aka open circuit) and indirect (aka closed circuit) cooling towers. In the cooling process, a small amount of the water is evaporated to reduce the temperature of the circulating water stream. Cooling towers recycle the water used in the cooling system. However, due to evaporation, the mineral concentration (blue colored in the bottom of Fig. 4.6) in water increases with every recycling round.

The primary mechanism of a recirculating cooling tower is to reuse the same water by circulating the water through heat exchangers, cooling canals, or cooling towers as depicted in Fig. 4.6. Besides the evaporation system, direct heat exchange using air circulation is also used as a cooling mechanism. Although the primary mechanism is relatively straightforward, the associated heat transfer equipment varies widely in cost and complexity.

4.4.3.1 Smart cooling tower systems

Cooling towers consume more than 20% of the overall sector energy usage. Hence, a smart cooling tower should be energy- and cost-efficient.

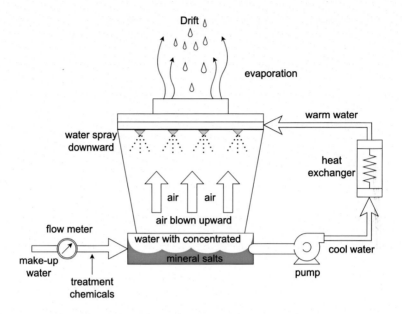

Figure 4.6 The cooling tower mechanism.

To achieve this goal, researchers analyze and synthesize the cooling water systems using the Kim and Smith design (KSD) method [109]. Specifically, the principle of the water sources diagram (WSD) procedure is employed to expand the heat exchanger network. As a result, the load and cost of the cooling tower are reduced.

4.4.3.2 Smart solutions for cooling tower issues

Cooling tower issues can incur higher costs and risks. Production constraints, unsuitable water quality, and hydrocarbon leaks are challenges in designing and maintaining cooling towers. The cooling tower monitoring and analysis solutions replace infrequent manual readings with online insight into the health and performance of cooling towers. To establish and function the cooling tower monitoring system, a smart digital architecture consisting of cutting edge technologies is necessary, which is presented in Fig. 4.7. The entire cooling tower system can be envisioned with modern computing technologies like edge, fog, and cloud computing, machine learning, from data generation to operational performance. The online insight detects and addresses issues that can lead to lost cooling capacity, asset damage, or higher energy usage. The following sections briefly describe some smart solutions implemented in the industry.

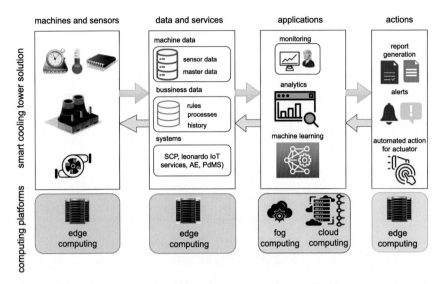

Figure 4.7 Smart cooling tower system architecture utilizing smart sensors, edge, fog, and cloud computing systems.

Water quality of cooling tower: The makeup and blowdown cycles considerably impact the cooling efficiency and water quality. On the one hand, low cycle rates can lead to poor water quality, which causes fouling, corrosion, and potential health safety issues for personnel. On the other hand, high cycle rates can cause excessive wear on equipment and higher chemical and makeup water usage and costs. Monitoring solutions that allow conductivity-based water quality control can optimize the cycle rate, minimize operating costs, and ensure good water quality.

Fan and supply pump vibration: Bearing failure or mechanical seal damage can lead to pump failure or fan trips. In addition, shutdowns of cooling tower fans and supply pumps can disrupt throughput and decrease cooling capacity. An integrated solution [110] can detect premature bearing wear signals early and prevent slowdowns and shutdowns.

Hydrocarbon leakage: A leakage from heat exchanger tubes can contaminate the circulating water that will be released into the environment. As a result, immediate corrective actions in leakage are essential. Monitoring the water's surface to detect any possible liquid hydrocarbons can help take early corrective actions [110].

Heat transfer efficiency: Sediments in heat exchanger tubes can reduce their heat transfer efficiency. Higher concentrations of minerals in circulating water caused by evaporations increase the sediments on heat exchanger surfaces. In addition, corrosion of equipment can reduce heat transfer. Therefore, effective monitoring of scaling and corrosion can increase heat transfer efficiency and lower energy costs. Moreover, the appropriate number of cooling tower units required to meet cooling water demands can be suggested based on the ambient conditions.

4.4.3.3 Artificial intelligence & IoT in cooling towers

Artificial intelligence technology, specifically machine learning algorithms, can play a crucial role in improving the cooling towers' performance. With the embedded weather sensors (*e.g.*, temperature, humidity, wind flow rate), cooling towers can predict the weather and adjust the cooling condition for that particular scenario. Furthermore, adjusting pumps' and fans' speeds and directions can optimize the cooling environment. Thus, consuming less energy by effective decision-making using machine learning algorithms improves the efficiency of the cooling towers in terms of energy consumption.

In a typical industrial scenario, the computing need for data analysis and decision-making is performed using cloud data center resources that may be located in a distant place. Here, the network latency issue for transferring data and getting the response back can create some trouble for real-time scenarios. Additionally, the security and cost of using cloud resources can be higher considering network bandwidth utilization. As such, private edge and fog computing platforms (located near or on-premise of industrial areas) can solve the security and latency issues. Fig. 4.7 represents the system architecture of a smart cooling tower system that works efficiently for both edge and cloud computing systems depending on different network conditions.

4.4.4 Smart boilers

Boilers that are designed as closed vessels are used to heat the water and convert it to steam. Heating the water will increase the pressure and eventually change its phase to gas.

4.4.4.1 IoT enabled boilers for refinery systems

The emerging IoT technology with Industry 4.0 revolution changed various manual operations into automation. Hence, the smartness of modern boilers mainly lies with energy efficiency and automated boiling operations that can be achieved by utilizing IoT and other state-of-the-art computing technologies (*e.g.*, edge computing, fog computing, high-performance computing). In addition to automation, a smart boiler system can be cost-efficient and more robust in terms of quality of service. Although the existing O&G industry uses Distributed Control System or Programmable Logic Controller (DCS/PLC) in refinery plants, the controllable parameters (*e.g.*, temperature, humidity, pressure) require well-management system for completing any real-time nature tasks without any accidents.

Data mining techniques can be used to develop a smart simulation and automated system for boilers in refineries. Specifically, these techniques are used to perform modeling and pattern discovery for collecting data from a thermal power plant [111]. The boilers are equipped with sensors and connected to an IoT application at a distant location. The smart sensors work with a particular pattern considering a specific situation that could develop while the plant is processing a particular product. Hence, it is possible to identify the behavior of a boiler usage from the sensor data; then, actuators can be utilized accordingly to improve energy efficiency and predictive maintenance.

4.4.4.2 Case study: Optimal design of heat exchanger using a combination of machine learning and computational fluid dynamics (CFD) methods

In the oil and gas industry, there are often severe restrictions on the volume and weight of the equipment. Also, the volume and weight optimization procedure need accurate thermal-hydraulic correlations. Computational Fluid Dynamics (CFD) can predict such thermal-hydraulic correlations. However, direct optimization is usually not feasible due to many design variables, constraints, and the computing cost of each function evaluation. To address these problems, researchers propose to combine predictive CFD simulations with adaptive sampling and automated correlation building methodologies based on machine learning theory [112].

They hypothesize that correlation accuracy and validity range can be increased simultaneously, with reasonable computing effort, by leveraging publicly available experimental work in conjunction with new, adaptively sampled simulation points. The presented methodology applies to a wide range of multivariate design problems where direct optimization with CFD is infeasible [112].

The overall correlation building and verification process are shown in Fig. 4.8. First, an initial database of experimental work from the open literature is used as the input of the ML model. Next, a correlation is generated and improved through an adaptive sampling of data points from the CFD

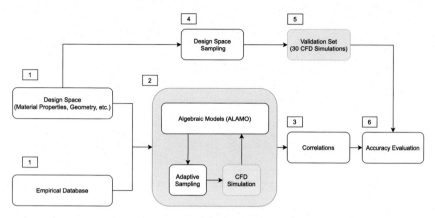

Figure 4.8 Process for correlation development, testing and benchmarking. Color boxes indicate where CFD simulations are employed [112].

simulator. Finally, the correlation is evaluated using a separate validation set. The validation set consists of 30 CFD simulations selected through a sampling method from the design space.

4.4.5 Smartness in distillate hydrotreater, fluid catalytic cracker and alkylation

4.4.5.1 Distillate hydrotreater

The decreasing availability of low-sulfur crude oils has led to more high-sulfur crude oils processing. For this reason, the number of hydrotreating processing units installed in refineries is increasing [113]. Hydrotreating as depicted in Fig. 4.9 improves the quality of oils by removing sulfur and nitrogen. In addition, it increases the hydrogen content by saturating olefins and some of the aromatics. The process is performed using catalysts to react hydrogen at high pressures with the feedstocks. It is sometimes called hydrotreating, hydrodesulfurization (HDS), or hydrodenitrogenation (HDN).

4.4.6 Fluid catalytic cracker (FCC)

Thermal cracking operations consider the hydrocarbon boiling point and heat it to perform the cracking. Recently, chemical catalyst utilization started to play a significant role in successfully refining petroleum. The unit that performs the hydrocarbon cracking using a fluid catalyst is the Fluid Catalytic Cracker Unit (FCCU) and is generally found in complex refineries as a secondary conversion unit. The working procedure of a FCCU

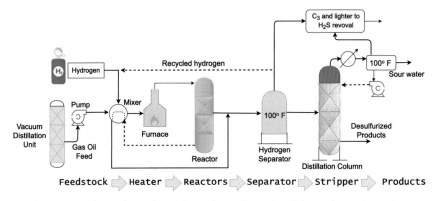

Figure 4.9 A typical hydrotreater process flow that starts with heater, reactor, separator, and finally go through stripper. The main purpose of this process is to remove sulfur and nitrogen from crude oil.

is presented in Fig. 4.10. The purpose of FCCU is to produce additional gasoline mainly from the gas oils that are produced in various distillation units [114].

As depicted in Fig. 4.10 at the reactor's entrance (referred to as the riser), a fluidized-bed (or fluid-bed) of catalyst particles is brought into contact with the gas oil feed and injected steam. When the heated catalyst particles from the regenerator unit contact the feed gas oil in the riser, the gas oil vapors and the catalyst particles migrate higher in the reactor, and cracking begins. The temperature of the catalyst particles decreases as the gas oil evaporates, and endothermic cracking events go higher. Additionally, cracking processes deposit a substantial quantity of coke on catalysts, resulting in the catalyst's deactivation. After steam stripping away the adsorbed hydrocarbons, the coked catalyst is transferred to the regeneration unit, where the coke is burned off with air. The heat generated by the coke deposit burns off the catalyst particles, which are then reintroduced to the riser to continue the cycle. Burning the rejected carbon (coke) in the regenerator provides the energy required for cracking with slight loss, enhancing the process's thermal efficiency. After being separated from the catalyst particles in the top portion of the reactor, the cracking products are transferred to the fractionator for recovery.

The cracking reactions begin in the reactor with the formation of carbocations. The subsequent ionic chain reactions produce branched alkanes and aromatic compounds comprising crackate (cracked gasoline with a high octane number), olefins, cycle oils, and slurry oil sent to the fractionator as

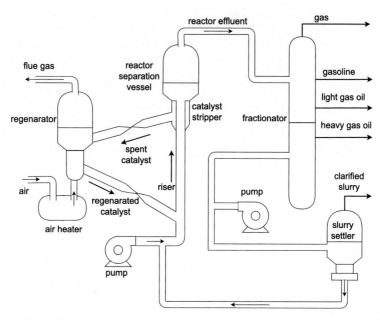

Figure 4.10 The fluid catalytic cracker process diagram. Catalytic cracking consists of three major processes namely Reaction, Regeneration, and Fractionation.

presented in Fig. 4.10. Coke, a carbon–rich byproduct of catalytic cracking, accumulates on the surfaces of catalysts and obstructs the active sites. FCC is a carbon rejection process because the coke formed on the catalyst surface and subsequently burnt off for heat is rich in carbon, enabling the process to produce vast amounts of a light distillate (crackate) without adding hydrogen.

4.4.7 Alkylation

Alkylation is a chemical process of producing high-octane gasoline. Specifically, the light hydrocarbons are sent to a reactor to generate a composition of heavier hydrocarbons. A block diagram of alkylation process is presented in Fig. 4.11. Olefins like propylene and butylene, as well as isoparaffins like isobutane, make up light hydrocarbons. According to Fig. 4.11, these compounds are put into a reactor, where they interact with sulfuric acid or hydrofluoric acid catalyst to generate a heavier hydrocarbon mixture. The liquid part of this combination, known as alkylate, is mostly made up of isooctane, a chemical that gives gasoline exceptional antiknock properties.

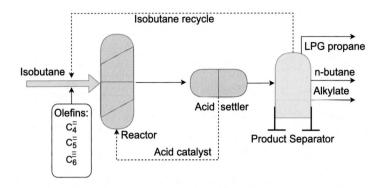

Figure 4.11 Block diagram of alkylation process.

Alkylation units were first deployed in petroleum refineries in the 1930s, but the process became vital during World War II when aviation fuel was in high demand. To boost a refinery's supply of vehicle gasoline, it is currently combined with fractional distillation, catalytic cracking, and isomerization.

4.4.8 Smart IoT based predictive maintenance

Unplanned downtime can incur a high cost to oil and gas companies. For example, it has been reported that 3.65 days of unplanned downtime per year can cost oil and gas companies $5.037 million per year [115]. In addition, the average unplanned downtime for an offshore oil and gas company is 27 days, which can cost $38 million [115].

Oil and gas companies search for new, more efficient maintenance methods to eliminate the risk of unexpected equipment failures and maximize return on assets. A smart solution of predictive maintenance using advanced computing technologies is demonstrated in Fig. 4.12. Why predictive maintenance driven by IIoT (industrial Internet of Things) is worth considering can be understood by exploring the process of predictive maintenance. In a nutshell, IIoT-driven predictive maintenance leverages data fetched from equipment's sensors (*e.g.*, temperature, vibration, flow rate sensors) to identify anomalies in equipment behavior and forecast equipment failure in a specific timeframe. The simplified process looks as follows:

- Step 1. Collecting IoT data
- Step 2. Adding context
- Step 3. Searching for patterns
- Step 4. Creating predictive models

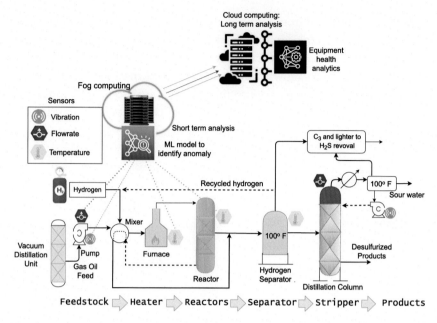

Figure 4.12 Predictive maintenance in hydro treating process. Fog computing plays the role of short term analysis and anomaly detection. Whereas, cloud computing is utilized for long term analysis of the equipment that is critical for hydro treating process.

One of the major reasons for unplanned downtime is poor maintenance of refinery equipment. Pumps and compressors are the critical types of equipment to maintain. These types of equipment are widely used in the downstream sector, mainly in oil distillation units, diesel hydrotreating units, fluid catalytic cracking units, and sulfur recovery units.

Production and environmental data with temperature, vibration, and flow rate data from sensors attached to the potential faulty points are used as the input of the predictive models, which enable refineries to predict whether a component is likely to fail long before a problem arises.

4.5 Distribution of end products

According to the taxonomy shown in Fig. 4.2, distribution is one of the major branches of downstream sector. Typically, refineries are located in the proximity of population centers to facilitate the marketing and distribution of final products. In general, final petroleum products (*e.g.*, gasoline, diesel, jet fuel) are distributed from refineries to end consumers via multiple channels (*e.g.*, pipelines, tank wagons, trucks). Hence, the distribution network that transfers petroleum products from the refineries to the end–users consists of pipelines, ships, railways, and trucks. As shown in Fig. 4.13, various means of transportation are used to move petroleum products from the refineries, ports, and large terminals to tremendously disperse markets. Two principal distributions are primary and secondary distribution.

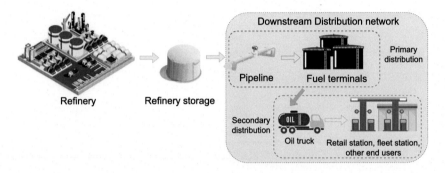

Figure 4.13 Downstream distribution network from refinery to end users.

In primary distribution, refined petroleum products are transferred from the refinery to the intermediate storage facility. Then, at a later stage, end products are withdrawn by different operators and delivered to end consumers. The pipeline is widely used among various transportation means due to its cost and latency efficiency. Barges, tank wagons, and sometimes tankers also perform primary distribution.

Secondary distribution, also called capillary distribution, involves distributing oil products to the end consumer via multiple channels (*i.e.*, tank wagons, barges, and trucks) to supply service stations, industries, ships, or airplanes.

4.5.1 Blockchain based supply chain management

Between the point of production and the ultimate user, a drop of oil or a cubic meter of gas passes through multiple owners, interacts with at least 30 counterparties, and includes 200 information solutions. All of these rounds are represented in each event participant's document flow. They often use disparate techniques to capture and store data, resulting in a plethora of mistakes and inconsistencies. All these transactions are compounded by the high degree of market price volatility and the complexity of compliance and regulation regimes. In addition, a large number of projects and sub-contractors, and workers with highly varying educational levels and skills led to the most complex document management system in the world.

Introducing blockchain technology is made possible due to the fact that the vast majority of data in the workflow of individual market partici-pants are shared. As a result, a single network can be constructed based on modern technologies to digitize and automate all interaction procedures, having reduced the number of errors, fraud, reconciliation, and settlements between counterparties. The blockchain enables the asset transfer contract to be integrated into the transaction database [116]. The execution of a contract is assured after it has been verified and deployed. All transactions against the ledger need network consensus, where the provenance of in-formation is not vulnerable to misunderstanding and is transparent to all participants. Transactions are final and cannot be altered.

Vertrax, a leading supply chain management solution provider has launched the first multi-cloud blockchain [117] solution built on IBM Blockchain Platform and deployed it on AWS to prevent supply chain dis-ruptions in bulk oil and gas transportation. The solution delivers flexibility, collaboration, and speed-to-market. Vertrax realized the need for oil and gas supply firms to unify their logistical systems. Oil and gas firms had to spend much money on operating expenses since many of their com-petitors were still utilizing antiquated technologies, such as paper record tracking. All supply chain players were expected to access real-time data regardless of their computing environment. Their ability to react rapidly in the face of various difficulties affecting the supply chain is facilitated by this system. As a result, using Vertrax's current and new oil and gas logistics systems and IoT sonar sensors, Chateaux [118], an IT service and solution provider, built a blockchain solution that provides end-to-end monitoring from the well to the residential end customer. The web app, created with IBM Blockchain, runs on several clouds, increasing user adoption and pro-viding built-in security and privacy features. Distributed ledger technology allows oil and gas bulk liquid distributors to acquire supply chain insights without compromising proprietary data.

4.6 Safety of refinery facilities

Downstream storage systems and processes have complex challenges not only with their operation but also in their development. Safety and system integrity are particularly important as optimization and efficiency during the design phase. This requires careful consideration of various safety components to ensure worker safety and long-term operational efficiency with little downtime.

Pressure relief valves (PRVs) and pressure safety valves (PSVs) are important examples, improving the system's safety by releasing excess pressure buildup before it can damage storage tanks and pipelines. In the case of PRVs, the valve is typically spring-operated, opening when internal pressures exceed spring pressure. PSVs, on the other hand, tend to be manually operated and used when an emergency strikes. Therefore, more advanced relief valves like vacuum pressure safety valves (VPSVs) and pilot-operated relief valves (PORVs) may be necessary for more challenging scenarios.

Aside from PRVs and PSVs, engineers have other tools for regulating pressures at a safe level. For example, thermal relief valves are installed when thermal expansion could devastate equipment. Pressure regulators can also be added inline, typically run open at the front end of a system to reduce the effects of higher upstream pressure. Backpressure regulators are similar; however, they typically run closed at the back end of a system to keep backpressure in check and aid in the drawing off of fluid or gas.

4.6.1 Refinery process safety

Process safety is a disciplined framework for managing the integrity of operating systems and processes that handle hazardous substances. It relies on design principles, engineering, and operating and maintenance practices. Specifically, it deals with the prevention and control of events that have the potential to release hazardous materials and energy.

The terms 'process safety' and 'asset integrity' are both used throughout the petroleum industry, often synonymously. For the oil and gas industry, the emphasis on process safety and asset integrity is to prevent unplanned releases resulting in significant incidents. A hazardous release typically initiates a major incident. In addition, a structural failure or loss of stability could also lead to a major incident.

4.6.2 Refinery equipment safety

Petroleum refineries consist of various types of equipment and machines. For instance, pressure vessels, fire heaters, piping systems, heat exchangers, storage tanks, pumps, and compressors perform different refining processes (*e.g.*, refining units, vessels, and columns for distillation). These types of equipment can malfunction or create hazardous substances that pollute the environment and workers. Hence, predictive maintenance of refinery equipment and continuous monitoring of hazardous areas is essential for the safety of a petroleum refinery.

All equipment in any industrial plant becomes vulnerable with time. The intense processes of an oil refinery, including extremes in temperature and pressure and caustic chemical compounds, often create a stressful environment for equipment. Hence malfunctions and outages in refinery equipment are inevitable. However, well-planned and executed maintenance programs can minimize the downtime and expenses caused by unexpected shutdowns. Therefore, a critical work program of regular and periodic plant maintenance and repair of equipment, materials, and systems is a vital division of any major refinery operation.

4.6.3 Smart predictive maintenance for refinery equipment safety

In the highly technical realm of an oil refinery, effective maintenance requires maintenance engineers and technical workers. They are mainly responsible for maximizing performance and cost savings. In addition, their efforts are to enhance the refinery's value—an objective sometimes called "asset integrity management (AIM)". However, manual maintenance work could be risky or life-threatening for workers. Hence, with advanced robotics and deep learning models, robot workers can be used to perform maintenance work even though computing support and machine learning models are needed to perform accurate and precise maintenance work.

Preventive maintenance is performed in regularly scheduled intervals based on the age and remaining life of the equipment. The frequency of inspection and maintenance is based on the probability and consequences of failure. Typically, preventive maintenance work is planned and completed when equipment is shut down. The purpose is to eliminate unnecessary inspection and repair tasks and reduce maintenance costs. Examples of preventive maintenance may include simple tasks such as cleaning, tightening bolts, changing oil, lubricating equipment, or adding parts such as filters to prevent impurities from contaminating products. Predictive maintenance is a continuous process based on the current condition of the equipment. Sometimes predictive maintenance is called condition–based monitoring. The purpose of predictive maintenance is to prevent unscheduled shutdowns by predicting damage and failures before they occur. This can be done through inspection methods or by utilizing sensors to collect data and measure the current condition of equipment during operation. Predictive maintenance uses measurements, signal processing, and oil and gas engineering to estimate the useful lifetime of the equipment. This type of maintenance supplements preventive maintenance, and it involves close collaboration with equipment vendors. Although installing predictive maintenance measures can be costly, such an investment shows an overall reduction in operational downtime.

4.6.4 Refinery workers safety

The petroleum refinery industry follows many safety rules, regulations, and periodic awareness events to ensure the safety of the workers. The workers' safety is divided into two categories based on the damage that happened to the worker: short term safety (*e.g.*, using protective gear, available first aid box, emergency rescue plans) and long term safety (*e.g.*, periodic health checkups, taking measures to reduce toxic gas emission).

Among various safety measures, one of the most important pieces of equipment to ensure workers' safety is Personal Protective Equipment (PPE) that workers need to wear during the operational activity in a refinery. The PPE mainly provides smartness in this area. The smart PPE includes a helmet with a smart camera, a bodysuit with different sensors, wearable watches, and smart shoes that is presented in Fig. 4.14.

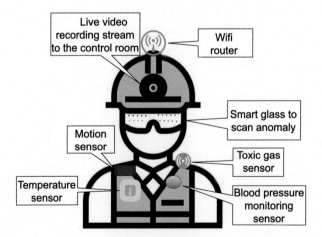

Figure 4.14 Refinery workers safety with smart personal protective equipment (PPE). Wireless sensor network and advanced edge computing friendly computing resources are enabled for real time support to ensure workers safety.

4.6.4.1 Chips in smart tech helmets

GPS chips used in smart helmets presented in Fig. 4.14 enable the supervisors to track the worker's location at the worksite in real-time. Furthermore, smart helmets are equipped with multiple sensors to collect information and protect workers. A smart helmet can detect impact, free falls, temperature, humidity and other senses. Moreover, helmets are programmable. They can be programmed to warn the user about the detected hazards nearby. The location information can assist in preventing accidents. In addition, traditional safety glasses can be replaced by smart eyewear in risky areas. Hence, displaying data inside the smart eyewear enables the wearer to monitor any data changes continuously.

4.6.5 Environmental impacts of oil refineries

4.6.5.1 Air pollution

In recent decades, due to the rapid development of the industry, a large amount of pollutants has entered the earth's atmosphere. However, an approach to reducing pollutant emission is to predict the pollutant dispersion around its industrial source.

Hashemi et al. [119] employed the principles of Computational Fluid Dynamics (CFD) to simulate the distribution of gaseous pollutants emitted from twenty-three stacks of different units of Shiraz oil refinery. They considered CO, HC, SO2, and NO as pollutants in their work. Then, the authors show the pattern of pollutants dispersion around the oil refinery as its source of air pollution. In addition, the comparison between the CFD simulation results with the existing empirical data a good agreement between the measured data and those obtained from the CFD simulation.

Rahimi et al. [120] developed a 3D CFD model to predict the dispersion of gaseous pollutants released from different stacks in the Isfahan refinery in Iran. Then, they compare the results of their model with the experimental data obtained by measuring the CO_2 concentration inside and close to the refinery boundaries. The comparison shows the sufficient precision of model predictions.

4.6.5.2 Water pollution

Petroleum refineries produce lots of industrial wastage that can pollute the surrounding water bodies. Additionally, refineries produce wastewater that can be harmful to the environment. Hence, the analysis of chemical contamination in petroleum wastewater is essential. Accordingly, it has been discovered that the number of organic chemical contaminants in oil refinery wastewaters has risen from less than 20 to over 300, which is alarming for environmental pollution.

Potentially, refineries are major source of ground and surface water pollution. Typically, deep-injection wells dispose of industrial wastewater that sometimes contaminates aquifers and groundwater. Finally, these wastes are monitored by the Safe Drinking Water Act (SDWA). Wastewater in refineries is sometimes highly contaminated. This contaminated water may be wastewaters from the desalting process, cooling towers, distillation units, or cracking units that may contain many hazardous wastes. This water is recycled and undergoes several treatment processes before being released into surface waters. Despite strict regulations for releasing wastewaters, contamination from past discharges may remain in surface water bodies.

4.6.5.3 Soil fertility

The impact of refinery processes on soil is generally less than on water and air. Due to past production practices, there might exist spills on the refinery property that should be treated. An effective way to clean up petroleum spills is natural bacteria which uses petroleum products as food. In addition, the refining processes could produce many residuals. Some residuals may be recycled during other processes. However, other residuals are collected and disposed of in landfills, or other facilities may recover them. Typically, soil contamination, including some hazardous wastes, spent catalysts or coke dust, tank bottoms, and sludges from the treatment processes, can occur from leaks and accidents or spills on- or off-site during the transport process.

4.7 Summary

Worldwide, the oil and gas industry recognizes the value of digitalization for their downstream processes. Nonetheless, they frequently have difficulties initiating and completing digitalization projects in their core operations, resulting in a quantifiable return on investment. Environmental pollution and worker safety also play a vital role in implementing advanced IoT and computing technologies. Given the technological advancement and the complexity of the downstream sector, refiners and petrochemical companies can realize the full potential of digitalization only if they take a planned, organized approach. Unplanned downtime costs global refiners $60 billion [121] in operating expenses each year due to ineffective maintenance methods. Oil refineries must thus have a suitable maintenance strategy in place to assist them in avoiding unplanned equipment failures. Refiners often plan maintenance for particular units or the refinery as a whole to change inspection coordination, repair operations, and equipment configuration. Refineries usually draw individual pieces of equipment into workstations for inspection and maintenance with little knowledge of their state. With the advent of smart devices (IoT Sensors), improved wireless mesh networks (Network), and device and asset management analytics (Augmented Intelligence), condition-based predictive maintenance is becoming more common. Due to the fact that refining activities dominate the downstream section of the oil and gas industry, it is critical to managing the storage and distribution of the industry's numerous products. Refineries can digitize their storage operations with the help of an IoT-level monitoring system. Fuel and other petroleum products are kept on a tank farm in enormous tanks. A big tank farm may include between 100 and 500 tanks. Keeping track of gasoline inventory in such a vast tank farm might involve many human site visits and inspections. The Internet of Things-based gasoline tank level monitoring technology lets refineries check the amount of fuel stored in tanks from remote locations. The sensors broadcast the level of tanks straight to a centralized dashboard accessible from any internet-connected device. This technology enables refineries to trace all petroleum products until they are supplied for commercial use. Hence, by utilizing smart IoT and computing technologies, the downstream of O&G can be benefited in many ways while keeping the surrounding environment safe and sound.

CHAPTER 5

Threats and side-effects of smart solutions in O&G industry
Smartness is not necessarily smart

5.1 Introduction and overview

The emerging IoT technologies, cloud-based computing systems, advanced wireless networks, and machine learning solutions have collectively enabled large-scale cyber-physical oil and gas systems. In particular, the wireless connectivity has changed the operational models, such that the most, if not the whole, oil and gas production workflow can be controlled remotely, by means of various sensors and actuators. Moreover, these technologies have remarkably improved the productivity and efficiency of the O&G industry. However, we should note that not everything about the paradigm shift in digitalization and smartness is bright! There are some downsides that must be taken into account too. In fact, the digital transformation and ubiquitous connectivity create vulnerabilities that can be harnessed by the intruders to perform cyber-attacks and threaten the production, distribution, and even safety of the oil and gas industry. As we have noticed in several recent instances, such as colonial pipeline [122], Amsterdam-Rotterdam-Antwerp (ARA) cyber-attack [123], and Norwegian energy company [124], malicious software systems (a.k.a. malware) have been able to takeover the control of a system and block its normal operation until the intruders' demand has been fulfilled. Indeed, these recent cyber-attacks have proven that cyber-attacks can be as harmful as physical attacks in terms of both implications and severity.

Addressing the security concerns of smart O&G industry requires advanced analysis and diagnosis of the whole system. However, specialized security solutions for smart oil and gas are complex to implement and are not available yet. As such, there is a lack or gap of security solutions of smart O&G that opened the doors for numerous cyber-attacks over the recent years and has posed serious threats to the societies. Considering the criticality of O&G in today's world plus the management and technical gaps

IoT for Smart Operations in the Oil and Gas Industry
https://doi.org/10.1016/B978-0-32-391151-1.00014-9

exist across different sectors of this industry, comprehensive security solutions are needed to be explored for this industry as part of the Industry 4.0 revolution.

One of the most vulnerable sectors of O&G is the growing network of connected "things" (IoT devices) being deployed across the upstream, midstream, and downstream sectors. Smart sensors generate a humongous amount of real-time and real-world data that have to be processed to achieve operational efficiency, such as predictive maintenance and safety of the on-premise workers. Connected cameras with object tracking capability, geofencing perimeter protection solutions, third-party intrusion, and other access control systems can be attacked to breach the physical security and/or safety of the O&G infrastructure.

Accordingly, this chapter of the book is dedicated to the side-effects of smart O&G industries and the ways to mitigate such side-effects. In the rest of this chapter, we survey all the areas in a smart O&G that can potentially raise a security threat. Then, we will explore solutions to handle these threats.

5.2 Taxonomy of cyber-threats and side-effects in the smart O&G industry

To categorically explore various pitfalls of smart solutions in the O&G industry, we develop a taxonomy that is presented in Fig. 5.1. In this taxonomy, we divide the potential downsides of smart solutions into two categories, namely *vulnerabilities* and *side-effects*. The vulnerability part explores the *cyber-threats* and the problems caused by *device incompatibilities* in a smart O&G system, especially, addressing software, hardware, infrastructure, and data related vulnerabilities in the oil an gas industry. In contrast, the side-effect category focuses on the problems arise due to the *interaction* with the smart solutions (*e.g.*, human-machine and machine-machine interactions) and *biases* in a smart system.

Fig. 5.1 categorizes various downsides of smart solutions implemented or will be implemented in near future. This taxonomy, is the road map of this chapter that helps to keep track of complex smart solutions and link the corresponding consequences. Hence, we traverse the taxonomy in the following sections to understand the scope of the downsides of smart solutions.

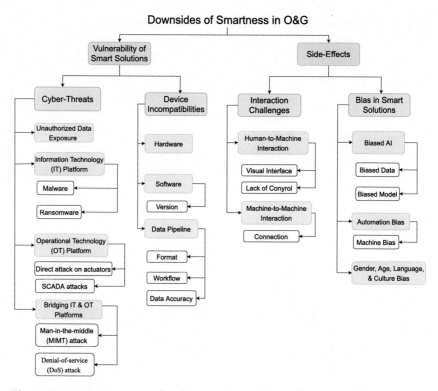

Figure 5.1 A taxonomy reflecting the downsides of smart solutions implemented with advanced technology is organized using box flow-chart form. The main three levels are colored in orange, blue, and yellow. The white boxes represent different types (examples) of its parent box.

5.3 Vulnerabilities caused by the interplay of informational and operational technologies

The technical operations of a smart oil and gas industry are divided into two major platforms: Information Technology (IT) and Operational Technology (OT). Fig. 5.2 demonstrates the overview of IT and OT components of an oil and gas company. As we can see in this figure, the IT component mostly deals with the flow of data and information within a corporate, whereas the OT component deals with the operation of physical processes of oil and gas production and the machines used to perform them. Cybercriminals mainly target IT and OT platforms to fulfill their demands. Conventionally, the IT component has been more vulnerable than the OT platforms as IT has many open windows (*e.g.*, operating systems, email servers, direct communication applications) that are suitable for the attackers. On the other hand, the OT platforms mainly deal with the direct oil and gas extraction and processing operations that have less accessibility from the outside world. Importantly, the intersection between IT and OT platforms often becomes the target for cyber-threats that has to be adequately addressed by the system architects. In addition, more recently with the prevalence of IoT, smart solutions using advanced computing technologies are creating accessible doors to OT platforms too. Therefore, we investigate the scope of the vulnerabilities in these two platforms and their intersection.

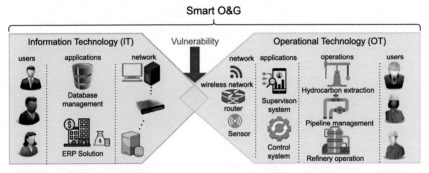

Figure 5.2 Information technology (IT) and operational technology (OT) platforms of a smart oil and gas company that operates using different networks to run the entire operation of smart O&G industry. The IT platform is significantly related to business applications and financial side of O&G, whereas OT platform directly involves with oil or gas extraction and production operations. Both IT and OT platform is connected at some point that creates the sweet spot for cyber-attackers to penetrate into the whole systems.

The OT platform consists of technologies that are directly involved in producing petroleum end products. The operations include extraction, refinery, pipeline, production, control systems, and supervision systems. On the other hand, the IT entity of oil and gas mainly deals with finance, database management, digital asset management, and other business activities using various computing platforms, and communication protocols. Here, the OT entity produces the petroleum end products, whereas, the IT entity creates business opportunities and financial policies leveraging output from the OT entity. Considering the activities of the IT entity, the interaction with outside network from O&G companies internal network is significantly high in comparison with OT entity. This interaction makes the loophole for a petroleum company to become victim of ransomware and other forms of cyber-attacks.

Traditionally, OT was an 'air-gapped' environment, meaning that it was not connected to external networks or digital technologies. Traditional OT has been relying on computers for several decades to monitor or change the physical state of a system, such as the use of SCADA systems to control train traffic on a rail system. In typical OT, the security is mostly composed of straightforward physical tasks, including making sure that a machine repeats the same task correctly, an assembly line continues functioning. Nonetheless, the rise of Industry 4.0 over the past few years has changed the landscape of traditional OT. Companies have begun implementing new digital solutions in their networks to increase the automation via adding "smart devices" that can capture data more efficiently and have network connectivity. As a result of this connectivity, and to process/analyze the OT data upon generation, the IT and OT systems became interconnected. Although this paradigm shift in the technology (referred to as *IT-OT Convergence* [125,126]) has created new opportunities and unlocked new use-cases, it has also introduced a scope for cybersecurity threats. For instance, Colonial Pipeline's attack [127] highlights how improper password management may damage the country's biggest gasoline pipeline. The hackers discovered the password for a VPN account that was no longer used but still functional. Oil and gas businesses should implement stringent cybersecurity measures, including personnel training, in light of this danger. Though not the first cyberattack against an industrial environment, STUXNET [128] was the first industrial control system (ICS) specific assault to get widespread notice. STUXNET is a harmful computer worm suspected of wreaking havoc on Iran's nuclear program, destroying over 20% of the country's nuclear centrifuges. Since then, cyber-attacks on industrial organizations have increased steadily, impacting a variety of sectors, including power grids (Industroyer), energy (Black Energy), petrochemicals (Havex), and oil & gas (TRISIS). Hackers are breaking into industrial networks to shut down machinery, demanding ransom, and stealing data, among other things. Operational Technology is often used to refer to the hardware and software that monitor and control the physical components of an industrial network (OT).

5.4 Cyber threats in smart O&G industry

The problem of the O&G industry is that it operates on systems that were not made with any kind of network connectivity in mind. For example, plants were never intended to be connected to networks, however, with the ongoing digital revolution, now they are. This can create dangerous situation, since a cyber-attack in such a system can sabotage the operations and can result in loss of lives too.

In general, the O&G industry trails behind in the matters of cybersecurity. Many companies still are not investing enough in solid cybersecurity systems, in spite of it being crucial for the industry's survival. Some of the security concerns that the O&G industry faces are mentioned in the rest of this section.

5.4.1 Vulnerabilities of sensitive data

Industrial IoT devices (sensors and actuators) as well as IoT platforms require strong security measures to protect their sensitive data. Today, oil and gas companies analyze such sensitive data from a variety of sources. These data sources include:

- Historical oil & gas exploration, delivery, and pricing data
- Demographic data
- Response data from job postings
- Web browsing patterns (on informational web sites)
- Social media
- Traditional enterprise data from operational systems
- Data from sensors during oil and gas drilling exploration, production, transportation, and refining

The above-mentioned items are sensitive data for any private company. Due to huge competition in the oil and gas market, various private data of one company can be significantly valuable for the rival companies. Therefore, these private data are becoming center of attention for the unethical hacker community.

5.4.2 Vulnerabilities of smart systems

In the previous chapters, we discussed the smart solutions in the upstream, midstream, and downstream of the oil and gas industry. Although these are a promising and futuristic solutions that could benefit the overall oil and gas industry, the vulnerabilities of these solutions should be also considered. The ways that a smart solution could experience a deficiency or be damaged are as follows:

Bias in the training of a machine learning technique: The performance of a machine learning model completely depends on choosing the appropriate training dataset. A dataset that is curated by selecting particular types of instances more than others is considered a biased dataset. For example, in visual methods for pipeline leakage detection by drones, if the image dataset used for training the model mostly includes sunny weather condition, the trained model would not perform well under the rainy or snowy weather conditions. Another example, in predictive maintenance, a model might have been trained for a specific brand of the equipment working in a certain conditions. As a result, this model could not generalize well for unseen data, thus, it would not be enough accurate to be used as a predictive maintenance model.

Uncertainty exists in the machine learning model: There is an intrinsic uncertainty in machine learning models which makes them vulnerable. That is, there is always a possibility of getting false-positive/negative result from the model output that may lead to a disaster in the worst case scenario. As an example, a smart fire detection system that fails to detect the fire in a refinery plant could quickly result in a huge damage.

Failure in the workflow of a smart application: Smart applications often are composed of multiple steps that collectively form a directed acyclic graph, known as DAG. For instance, detecting worker's faces in an oil rig requires capturing videos, extracting frames, and then process each frame individually. Interrupting somewhere inside such a smart application workflow can potentially result in the failure of the entire application, thus, making it vulnerable. A failure in sending the command to the actuator can disable the whole workflow, hence, losing the control over the pipeline.

5.4.3 Malware and vulnerability of information technology (IT)

Malware, short form of "malicious software" refers to any intrusive software developed by cybercriminals (often called "hackers") to steal data and damage or destroy computers and computer systems. Examples of common malware include viruses, worms, trojan viruses, spyware, adware, and ransomware. Recent malware attacks have exfiltrated data in mass amounts. In order to remove malware, one must be able to identify malicious actor(s) quickly. Fig. 5.3 shows different types of malware.

A *virus* is a malicious software attached to a document or file that supports macros to execute its code and spread from a host to another host.

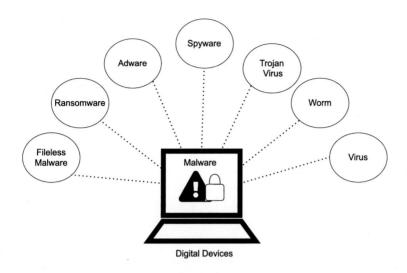

Figure 5.3 The types of malware threats in IT security.

Once downloaded, the virus will lay dormant until the file is opened. Viruses are designed to disrupt a system's ability to operate. As a result, viruses can cause operational disruption and data-loss.

A *worm* is a malicious software that rapidly replicates and spreads to any device within the network. Unlike viruses, worms do not need host programs to disseminate. A worm infects a device via a downloaded file or a network connection before it multiplies and disperses at an exponential rate. Like viruses, worms can severely disrupt the operations of a device and cause data loss.

Trojan viruses are disguised as helpful software programs. But once the user downloads it, the Trojan virus can gain access to sensitive data and then modify, block, or delete the data. This can be extremely harmful to the performance of the device. Unlike normal viruses and worms, Trojan viruses are not designed to self-replicate.

A *spyware* is a malicious software that runs secretly on a computer and reports back to a remote user. Rather than simply disrupting a device's operations, spyware targets sensitive information and can grant remote access to predators. Spyware is often used to steal financial or personal information. A specific type of spyware is a key-logger, which records your keystrokes to reveal passwords and personal information.

An *adware* is a malicious software used to collect data on your computer usage and provide appropriate advertisements to you. While adware is not always dangerous, in some cases adware can cause issues for your system. Adware can redirect your browser to unsafe sites, and it can even contain Trojan horses and spyware.

A *fileless malware* is a type of memory-resident malware. As the term suggests, it is a malware that operates from a victim's computer memory, not from files on the hard drive. Because there are no files to scan, it is harder to detect than traditional malware. It also makes forensics more difficult because the malware disappears when the victim computer is rebooted.

A *ransomware* is a malicious software that gains access to sensitive information within a system, encrypts that information so that the user cannot access it, and then demands a financial payout for the data to be released. Ransomware is commonly part of a phishing scam. The steps of a ransomware attack are shown in Fig. 5.4. As we can see, in such attacks, by clicking a disguised link, the user downloads the ransomware. The attacker proceeds to encrypt specific information that can only be opened by a mathematical key they know. When the attacker receives payment, the data is unlocked. Because this type of attack has become very popular over the past few years, particularly, in the O&G industry, we discuss it with more details in the next section.

5.4.3.1 Ransomware attack incidents

In a targeted attack, specific viruses serve multiple purposes: intrusion, data theft, dissemination, and more. Keeping a foothold in a victim's network is important to the threat actor. They need to continuously send their malware commands and receive data. DNS tunneling [129] is a process by which the DNS protocol is used to transfer data between the malware and the controller. Both email and cloud services can be used as the means of communication.

In the past, cyber criminals spread ransomware everywhere they could, often using spam botnets to try and hack as many computers as possible [130]. While it remains a serious threat to anyone who stores data on their device, ransomware has become an even greater threat as ransomware players target companies directly, with attacks that may have a major impact on their day-to-day operations. Some of the major ransomware attack incidents are as follows:

BitPaymer19: BitPaymer19 [130] is one of the dangerous ransomware family that targeted a U.S. company specialized in oil well drilling services. Actors behind BitPaymer usually use spear phishing to infect their targets with initial malware before moving laterally and compromising the network

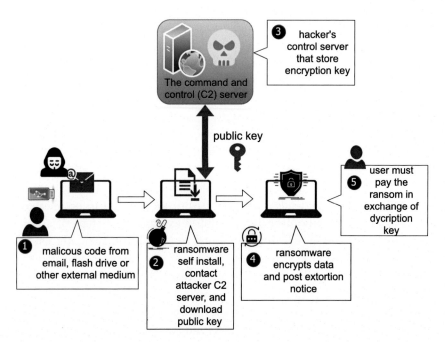

Figure 5.4 The anatomy of ransomware from start to end. The ransomware client enters the IT platform through malicious email or other external mediums. The client then communicates with hacker's command and control server to download the encryption key. The user's data encrypted by the ransomware client, and finally extortion notice is sent.

further. They plan the ransomware in different locations, and depending upon the absence of IT people (*e.g.*, they can plan the ransomware on weekends and holidays).

APT33: A famous target actor group focuses on targeting the oil industry and its supply chains. APT33 has shown particular interest in the aviation sector that are active in both military and commercial capacities, as well as in the energy sector organizations with links to petrochemical production. APT33 has also affected European and Asian energy firms. A few oil companies' IP addresses interacted with the C&C timesync.com website, which hosted a Powerton C&C server from October to December 2018 and in 2019. In November and December 2019, a database server operated by a European oil company in India interacted with a Powerton C&C server used by APT33 for at least three weeks. It has been found that APT33 was possibly infiltrated by a major UK-based firm providing specialized services to oil refineries and petrochemical facilities in Fall 2018.

APT33's best-known intrusion strategy was by emails using social engineering. It has been using the same type of lure for many years: a spearphishing e-mail containing a work opening bid that may seem quite valid. Campaigns also targeted hiring process in the oil and aviation sectors [131]. The email includes a path to the malicious ".hta" code. The ".hta" file may attempt to download a PowerShell script which can download additional APT33 malware to allow the group to stay in the target's network.

Ryuk: Another Ransomware attack occurred with the Mexican state-owned oil firm Premix that was hit by a ransomware attack which halted their critical operation and forced them to disconnect their network from the Internet and back up their critical information from their storage systems. As per the reports by Bloomberg, Premix servers were infected by Ryuk, a ransomware strain that generally gets distributed via email phishing campaigns or botnets [132].

5.4.4 Vulnerabilities in operational technology (OT) protocols

Even the conventional industrial control systems in the oil and gas sites are/were equipped with a tremendous number of devices from a wide range of vendors. However, because of isolation of these systems and lack of connectivity to the Internet, the industrial control systems were not considered high-risk for cyberattacks. Nonetheless, modern control systems that utilize monopolized technology platforms, enjoy IT-OT data sharing, and cloud-based services are more prone to cyberattacks. Hence, security companies are currently focusing on developing solutions to protect the whole IT-OT systems that consist of various physical assets and industrial processes. Both IT and OT need to be part of a comprehensive digital security strategy that is managed by a collaborative IT/OT team to accomplish end-to-end oil and gas production security.

5.4.5 Improving the security of the OT platforms

In Industry 4.0, the IT systems collect a significant amount of data from the OT platform and infer valuable insights about petroleum production efficiency. Hence, IT and OT systems communication networks need to be strengthened to avoid potential cybercrimes. The importance of enabling network isolation is vital and it also puts constraints on the accessibility of the OT systems to perform network anomaly detection. Implementing early attack detection systems, such as intrusion detection systems (IDS) and intrusion prevention system (IPS), in the OT platform reinforces the security of the O&G operations. However, sometimes falsely identifying authorized personnel as an intruder (*i.e.*, false positives) can cause network congestion, physical damage, or safety issues in the facility.

A security survey, in 2019, reflects that 87% of respondents did not correctly recognize the consequences of their security policy and strategies against cyber-threats. In this regard, we need a coherent security approach that moves beyond typical isolated security devices without compromising the production efficiency. A well-connected, broad, and automated security system can effectively monitor all the incoming traffic, and detect threats across the OT platform. Lastly, educating the OT users and administrators about the new systems, areas of exposure, and security measures is vital to ensure the utmost security of the OT platform for the oil and gas industry.

5.4.6 Data siphoning: vulnerabilities in IoT data transmission

Similar to an "eavesdropping" type of attacks, data siphoning is focused on the data being transmitted by an industrial IoT device, rather than an end-user. In this case, attackers eavesdrop on the network traffic traveling from the endpoint device back to the primary network to gather information they are not authorized to access. This particular type of attack is most concerning when the data that the IoT device is sending is sensitive and there is a chance that falls into the malicious users. In this case, it is important that the transmitted data be appropriately encrypted and protected against malicious accesses.

5.4.7 Vulnerabilities of IoT devices

IoT device hijacking occurs when a malicious actor takes control of an IoT device or sensor, often without the owner being aware that a breach has occurred. Depending on how "smart" the IoT devices are, the hijacking can vary in terms of how big of a risk or concern it can pose. If an endpoint or Industrial IoT (IIoT) device is compromised by a ransomware or malware, a malicious actor may be able to control the activity of the endpoint device. This is especially concerning if that endpoint or device has automated functionality, controlling manufacturing, or controlling the function of an internet-connected product in the field.

This can often happen if the IIoT devices are not properly updated. Typically it also may be the starting point for an attack that goes after the entire network by starting at an endpoint and using that device to gain access to centralized network. As many devices in manufacturing plants or within warehouses rely on old and legacy technologies, they may not be updatable and connecting them to the network create security loop wholes at the device level. Using a hardware-based VPN solution is often the only way to provide security to both the IoT device itself and to the data or information that it transmits.

5.5 Incompatible IoT devices

According to Fig. 5.1, incompatibility of IoT devices is one of the main sources of vulnerability in smart O&G industry. In fact, the main driving forces of a smart industry like oil and gas are the automated systems that receive inputs from various types of IoT devices and sensors, decide based on their internal business logic that functions via machine learning or statistical models, and then apply their decisions through different actuation operations. In practice, these connected IoT devices and sensors are produced and procured from different vendors, hence, are intrinsically heterogeneous. Such a heterogeneity is the source of incompatibility and has the potential to be misused by the cyber-attackers or create deficiencies during an emergency situation. Configuring connected and compatible IoT devices that can communicate smoothly are critical for efficient data transfer and provide real-time communication during emergency situations (*e.g.*, toxic gas detection).

Because procuring homogeneous and fully compatible IoT devices is hard or impossible to achieve, an alternative solution that is being explored is to design standard protocols that can facilitate efficient communication across all industrial IoT devices. That is why, the premier IT companies are working together to develop an unified protocol (known as the matter protocol [133]) that can support all the IoT devices available in the market. The matter is an emerging intelligent home inter-network protocol supported by Apple, Google, Amazon, and other major smart home players in the Connectivity Standards Alliance [133]. The matter is a unifying and IP-based connectivity protocol built on proven technologies to build reliable and secure IoT ecosystems.

We can consider three types and levels of incompatibility in a smart O&G and other smart industrial environments. These are namely, hardware level, software level, and data pipeline level incompatibilities that are explained in the next subsections.

5.5.1 Hardware-level incompatibility

Commercial off-the-shelf products are increasingly utilized in oil and gas to perform tasks that were originally undertaken by purpose-built equipment. Because of this general-purpose nature, they are often more prone to vulnerabilities and security risk, as opposed to traditional process control systems. Large-scale deployment and diversity of their usages in various domains dramatically increase the attack surface. For instance, in an attack scenario to an oil field, a smart real-time video camera for monitoring a hazardous area for anomaly detection could be hacked (as it has connectivity with internet), and at the same time an actuator valve can be released by hacking the controller system to release toxic gas or flammable gas. If all the sensors, actuators, cameras, and their supporting hardware produced by the same company utilize a compatible protocol to perform security check internally for any unusual behavior detection, then this attack could be avoided.

5.5.2 Software-level incompatibility

Software level incompatibility mainly refers to third-party software and outdated software leading to system glitches. In addition, third-party software sometimes carries malicious code that infects the internal systems. On the other hand, obsolete software systems need new updates to align with the operating system or firmware. Lack of updates or using incompatible updates that are not originally made for those software systems makes the systems prone to attacks.

The oil and gas companies purchase digital products assuming that the products are secure and can be integrated to the whole system. In most cases, they do not verify for incompatibility with the rest of the system. Moreover, the oil and gas companies sometimes even do not have the time, expert knowledge, or tools to affirm the software level incompatibility. That is how hackers exploit the flaws within the smart oil and gas and hack into the company's' internal networks.

With evolving IoT smartness, the industry leaders currently need new approaches to reduce cybersecurity risks throughout the petroleum supply chain process in both cases of buying software products or developing them. Software-level vulnerabilities have led to some of the most surprising cyber-attacks in the recent years. For instance, during the "NotPetya" cyber-attack in 2017, one energy company, banks, metro systems, and one of the world's largest container shipping companies were targeted by a malware. Interestingly, the malware was transmitted through the updating process of an accounting software package that was generally used by companies in Ukraine. The malware then outstretched to other systems, causing their systems to crash. Likewise, cyber-terrorists injected a customer-ready certified software with malware in the "SolarWinds" attack (2021). In both of these cases, cyber-criminals used software vulnerabilities to penetrate into the connected suppliers' systems. They setup back-doors that could be exploited to steal IP information later or install malware that could propagate throughout the customer systems.

5.5.3 Data pipeline incompatibility

Just as oil and gas move through a pipeline, so does data. Beyond protecting the physical security of oil fields, refineries, and distribution channels, companies must ensure that the data pipeline is secure across upstream, midstream, and downstream. IoT devices are becoming more prevalent in the oil sector as sensors and edge processing becomes more prevalent. The drawback to the expansion of digital technology in the oil and gas business is the flood of new cyber security vulnerabilities. Petabytes of the reservoir and operational data need to be transferred with a secure data pipeline and stored in cloud data lakes. Threats to nation-states vary from financial gain to espionage or sabotage. Recent cyber espionage cyber-attacks on the oil and gas sector include the cyber espionage organization APT34 or OilRig [134], masquerading as a researcher at Cambridge University to issue LinkedIn invitations and distribute malware on client systems in the UK. In addition, the Triton/Trisis hack by threat organization Xenotime began with a Saudi petrochemical complex, shutting down industrial safety systems, and then spread to power utilities in the United States and the Asia-Pacific area. Cybercriminals have also used ransomware to attack European oil and gas companies through phishing emails. Considering what has been revealed thus far about cyber events, threats, and adversaries, it is clear that the vulnerabilities, whether already existent or newly formed, stem from the need for data flow between the many functions of an oil & gas asset and organization. Data flow between field operations and information technology/operational technology systems and corporate information technology systems, or data centers, creates many security breach channels, and vulnerabilities [135] that need proper attention.

Secure data transfer solution: One secure data transfer solution is unidirectional security gateways, sometimes known as data diodes [136], which may assist provide secure, safe, and dependable one-way communication between two networks or domains. The data can only go in one route, away from the protected system, due to the physical restrictions imposed by these devices. This indicates that the data cannot be physically transmitted back into the protected system. They also offer high-security network interfaces for managing transfers and packet conversions across internal nonroutable interfaces. As a result, although the protected system may communicate information with others, it is entirely immune to cyber-attacks from the outside world.

The simplicity of this technique eliminates the need for continuing maintenance of firewall rules and worries about firewall software vulnerabilities, which are common in traditional network security systems. In addition, data diodes may be employed to provide high-security bidirectional communication connections. In this situation, separate data diodes are employed to restrict communications in both directions without allowing any paths for traditional protocol exchanges that may convey malware. Furthermore, communication may still be begun only on the network's high security or "trusted" side, reducing the possibility of data transmission being hijacked for malevolent reasons.

Oil and gas companies must think about security not in terms of "physical" versus "digital" but rather from end-to-end, with a corresponding plan, and ongoing monitoring and management. This approach needs to unify everyone in the value chain—suppliers, partners and even customers—around a singular standard security framework.

5.6 Blockchain to overcome cyber-threats in smart O&G

Although blockchain technology has become popular as the backbone of cryptocurrencies (*e.g.*, BitCoin, and Etherium), it is, in fact, a powerful security technique. It stores the data in the form of a linked list of "blocks" where each block maintains the previous block's hash value. Therefore, in the event of tampering with a block of data its hash value becomes inconsistent with the successor block, and the attack is identified. As a result, blockchain technology is being adopted in different parts of Industry 4.0 and smart O&G.

5.6.1 Blockchain-based control systems (SCADA)

The Industry 4.0 revolution has drastically changed how IT and OT technologies operate. Most OT platforms are mainly built on top of SCADA systems that collect data from smart IoT devices and transfer them to servers for analysis. This data acquisition approach is insecure and untrustworthy, which creates a loophole for cyber criminals. Therefore, blockchain–based security mechanisms based on edge and fog computing have been proposed to secure the data acquisition transactions of the SCADA systems. Fig. 5.5 presents the high-level overview of this mechanism where the collected sensor data are encrypted in data blocks before being processed on a cloud-based SCADA system. Each block keeps the hash of the current data and the hash value of the previous block. Then the block verification method involves all relay servers and the Data Aggregator (DA). For verification, the servers respond many times. If all parties agree that the block is genuine, the DA delivers the request to all servers. Finally, the DA successfully transmits the revised blockchain to the control center after adding the block. In a recent work [137], a more secure Industry 4.0-friendly consensus mechanism, and mining node selection mechanism have been proposed that are described in the next parts.

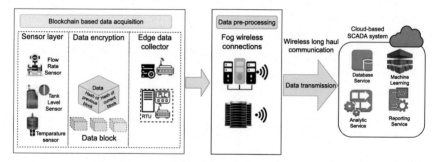

Figure 5.5 Blockchain based data transmission within end-to-end SCADA system of an oil and gas company. Blockchain enables encryption while transmitting the data for processing that increase the data security even data is hijacked while transmitting.

5.6.1.1 Consensus mechanism

At a high level, a consensus mechanism can be described as the process of adding newly published blocks in the blockchain after testing the validity by validators/miners to ensure the trust in the network. Over the years, numerous consensus mechanisms have been developed in several blockchain networks. Both public and private blockchain use some form of a consensus mechanism. Popular consensus mechanisms are PoW (Proof of Work), PoS (Proof of Stake), DPoS (Delegated Proof of Stake), PBFT (Practical Byzantine Fault Tolerance), PoA (Proof of Authority), and RAFT [138]. Each consensus mechanism has its own pros and cons. PoW, for example, imposes a high computing overhead and is biased to the wealthiest validators and thus unfair for new participants, Alternatively, DPoS is less decentralized and resilient. PBFT does not have anonymity, hence, only used for permissioned (*i.e.*, not publicly available) blockchains [139].

5.6.1.2 Mining node selection

In a blockchain network, a node is a computer that runs blockchain software and aids in the transmission of data. A laptop, phone, router, and so on are all nodes in the network. The nodes that contribute in performing the blockchain transactions and verifying them are known as "mining nodes". There is an option for any node in the blockchain network to be a mining node. Mining is a term used to describe the process of adding new transactions to a blockchain. The Data Aggregator (DA) edge server, as depicted in Fig. 5.5 collects data, processes it, and coordinates mining node selection and verification. The DA should be located within the internal network with the relay servers for minimum computing overhead and time. That is why, the fog servers in the data preprocessing phase of Fig. 5.5 are located within the internal network of DAs.

A customized mining node selection procedure has been proposed by the authors in [137]. First, the DA server requests data from the relay servers. After getting the measurement data from each relay, the DA generates a random number and broadcasts this random number. Relay servers then hash their measurement data and determine the random number appearance count from their respective hash values. Every server, including the DA, has the same counts in this stage. In the end, all servers vote for the relay server with the most random number of appearances. If all relay servers agree, the DA server chooses the highest count relay as the mining node for that cycle. However, suppose, that many relay servers have the same greatest count, or all have zero. In this case, the DA uses a cryptographically safe mechanism [140] to choose the mining node randomly.

5.6.2 Blockchain to enable trust across industrial IoT

One of the obstacles in the security of industrial Internet of Things (IIoT) is the trust issue. The traditional Public Key Infrastructure (PKI) architecture, which is based on a single root of trust, does not work well in this heterogeneous distributed IoT ecosystem that are potentially under distinct administrative domains. For this type of environment, a distributed trust model is required that can be built on top of existing trust domains and generate end-to-end trust across IoT devices without relying on a single root of trust. Therefore, providing a credit-based Blockchain with a built-in reputation system can be instrumental [141].

Another use of blockchain in the oil and gas industry might be to keep the credentials needed to operate safety-critical industrial equipment. Employee and contractor certifications, such as H2S training, first aid, welding, and so on, maybe securely recorded and stored on a company's blockchain network. All members may accomplish verification of credentials and standard operating procedures at any time by keeping such information in a blockchain network [142].

5.6.3 Blockchain for result verification in compute offloading

The smart oil and gas industry is equipped with numerous IoT devices that collectively generate huge amount of operational data every day. The smart IoT devices are not computationally powerful to process resource-hungry machine learning applications. Hence, these compute-intensive tasks can be offloaded to nearby fog datacenters for processing. In this case, computing resources of the fog systems need to be matched with the offloaded task. On the other hand, both the offloading party (in this case, oil and gas company) and the fog service provider have to build trust, via verifying the processing results, to ensure a reliable service offering.

Blockchain technology can be utilized for a secure offloading and for result verification. In [143], blockchain technology and smart contracts have been employed for result verification using the zero-knowledge proofs (ZKPs) concept. Smart contracts [144] are blockchain-based programs that execute when specific criteria are satisfied. They are often used to automate the implementation of an agreement so that all parties are instantly confident of the result. Also, the central concept of ZKPs is to convince someone that a statement is true without disclosing any underlying information.

In [143], as depicted in Fig. 5.6, a smart contract is implemented between the service consumer (*i.e.*, O&G company) and service provider (*i.e.*, cloud service provider). As we can see in this figure, the task offloading goes through the smart contract and verification steps to the service provider. Hence, the verification process allocates a blockchain-based cryptographic proof ("verifiable off-chain computation") with the offloaded task that testimony correct computation. Finally, when the result is sent back from the service provider (see the two-way service level agreement in Fig. 5.6), the cryptographic proof is also published. Then, the correctness of the computation can be verified on-chain using the smart contract conducted by the

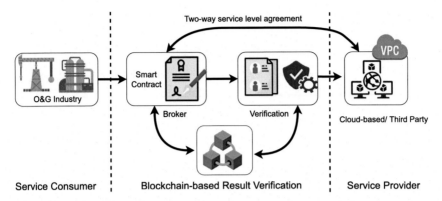

Figure 5.6 Blockchain-based secure result verification process for offloading tasks to third-party or a cloud service provider. Offloaded tasks (*e.g.*, ML-based prediction, reports, and automation) need verification after the process is complete. This can create a loophole for cyber criminals. Hence, blockchain technique is proposed with smart contracts to verify the offloaded workload of the smart oil and gas industry.

broker. This is one way to show how the integrity of an offloaded task's computing result can be ensured.

5.6.4 Aligning IT and OT to fill the security gaps

Cybersecurity incidents continue to show how loopholes in the IT, OT, or in the intersection of those can be exploited to cause substantial manufacturing disruption. Particularly, because of the linkage between IT and OT, an attack in one part of the organization can be like a fast-moving fire and affect other parts too. In addition, IT and OT systems are typically in lack of a unified management and they are maintained by separate teams with different goals and organizational structures. As such, attackers have more scope and opportunity to exploit the security holes.

Since the late 1960s, operational technology (OT) has been a part of the production process, utilities, and other industries. Although most OT devices were not connected to the outside networks, consumers thought the technology was "secure" from attacks for a long time. These devices are no longer air–gapped in today's IoT-based systems and that is why the risk of an attack has substantially increased, particularly on the OT part, which is more vulnerable due to lack of standard operating systems and other security measures.

To avoid such attacks, IT and OT teams must develop a unified, enterprise-wide security plan to eliminate blind spots. Combining IT and OT security, on the other hand, is not easy because OT networks are frequently larger and more complicated than IT networks. In addition, they contain assets, proprietary procedures, and processes that IT security tools and sometimes IT teams are unfamiliar with. Often, OT systems are embedded systems that function with nonstandard protocols and sometimes without operating systems or proprietary firmware that are not well-supported.

Conventional out-of-the-box IT network security devices are not effective in the industrial facilities. At the same time, basic cyber security approaches such as patching, encryption, and up-to-date anti-viruses are necessary for the OT environment. In addition, industrial cyber security systems require embedded machine learning and behavioral analytics to comprehend typical traffic patterns and detect suspicious activities.

Cyber security specialists recommend to break down silos between IT and OT and managing all cyber security under a single cyber security and risk team to effectively mitigate the risk. In this regard, the managers and security analysts of oil and gas plants should consider answering the following strategic questions:

- What are the devices, how are they connected, and how do they communicate with each other?
- What are the threats to the "crown jewel" IoT and industrial control systems (ICS) assets, and how to prioritize addressing them?
- Are there any IoT or ICS threats in the network, and if so, what is the strategy to address these threats quickly?
- How the existing resources (such as people, training, and tools) can be leveraged to help the operation team in centralizing the IT/OT security?

5.7 Risks of smart solutions in industrial IoT

As technology continues to advance and more of the world, including manufacturing plants and products themselves, becomes connected, understanding the risks associated with industrial IoT deployments is of paramount importance. Organizations consider launching a manufacturing or industrial IoT initiative, or connecting existing technology for automated and remote monitoring, will need to consider all of the potential risks and attack vectors associated with those decisions. The potential adversarial effects of smart solutions are discussed in the following subsection.

5.7.1 Human-machine interaction issues

The industrial IoT has improved a lot for connected devices to perform automated processing utilizing machinery. To ensure the efficiency of the production and/or safety of the workers conducting operations using various machines, different automated sensors and actuators (*e.g.*, video cameras, smartglasses, and automated valve with audio input/output) are in place to assist or replace the human worker.

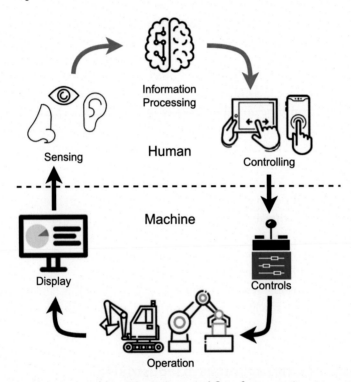

Figure 5.7 Human-machine interaction workflow from sensing to control operation.

Fig. 5.7 represent the human–machine interaction workflow separating the human side from the machine side. As depicted in this figure, human utilizes various senses (*e.g.*, sight, smell, hearing) to observe the results generated by the machine. Accordingly, human worker utilizes information processing to control or operate the machines. Then machines perform their operations, and generated results are displayed for the human interpretation. This entire cycle of operations is known as the human–machine interaction process.

However, the interacting machines could malfunction or not be user-friendly to perform the operation smoothly, which can be harmful to the production or to the onsite workers. Other than interacting with computing interface with input/output devices, sometimes human interactions are needed for conducting crucial industrial operation or, more importantly, to *override* the decision of a smart system. For instance, consider a drone or ROV that is tasked for automated surveying for oil and gas in a geopolitically-sensitive zone (*e.g.*, border areas). However, due to an inefficiency or bug in its algorithm, it may survey areas outside the appointed zone and does not let the operator to navigate the survey path. Such an interaction problem between human and machine can potentially lead to unintended political or military consequences.

5.7.2 Machine-to-machine interaction issues

In connected autonomous systems, machine-to-machine interactions occur to handle various tasks. In these systems, a series of actions are carried out automatically by multiple devices. If something goes wrong, it could be because: (A) the output is wrong (for example, an automated valve shut down because of a false anomaly detection, or an automatic door closing that traps onsite workers with a false alarm); or (B) the devices don't work well together, as we've already mentioned about.

Fig. 5.8 shows one scenario indicative the consequences of machine-to-machine interaction issue. Consider in a production environment pipeline maintenance, and connectivity with various machines need to be performed. Here a pipeline connected with a machine (*e.g.*, distillation unit) in an enclosed area need some maintenance. Accordingly, maintenance workers are working inside the enclosed area, when the event of fire hazard takes place. As a safety measure, the gas sensor detects the smoke (Step 2 in Fig. 5.8), start water sprinkler (Step 3), and sends alarm to controller. Due to water falling from sprinkler the power generator goes down (Step 4). To ensure the outside workers safety the controller sends the automated door to close (Step 5) immediately, ignoring the workers inside the area. In this scenario, the controller is unable to detect workers inside the facility and performs a safety action for outside workers that leads to a dangerous situation for the workers inside the facility. We can see that machine-to-

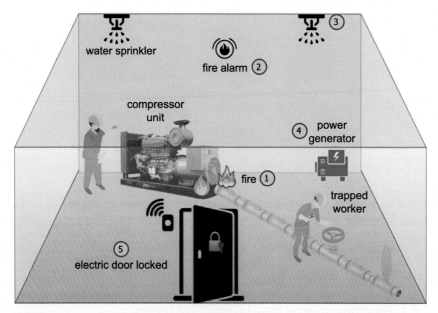

Figure 5.8 A small fire breakout accident occurs in a closed oil production area in a compressor unit. The fire alarm generates, and water sprinkler starts to sprinkle water that causes power failure in power generator that made the electric door locked. Unfortunately, workers were working on pipeline maintenance, and were trapped inside the facility due to door closure. Here, machine to machine interaction causes the safety issue of the onsite maintenance worker.

machine interaction issues can sometimes lead to situations that need to be addressed by through testing of a smart solution for the O&G industry.

5.8 Bias in smart O&G industry

Smart software solutions are prone to various biases that comes from human interference in developing various phases of the software [145,146]. Sometimes the biases can lead to accidents and hazardous incidents. Some popular examples of biases are gender bias, age bias, ethnic bias, physical challenge bias, among many others. Onsite staff (*e.g.*, worker, engineers, coordinators) utilize various software tools and simulations that can suffer from the above-mentioned biases. In the rest of this section, we focus on various types of bias and their potential impacts in the context of smart O&G industry and other contexts.

5.8.1 Biases caused by the artificial intelligence (AI) solutions

Although AI systems have been transformative in different areas, they are known to be biased in the following two ways:

- The model performs worse for certain inputs, due to insufficiencies of that type of input in the training dataset.
- The model replicates the existing bias in the training dataset.

The commercial facial recognition system is an example of AI bias due to insufficient training datasets. Researchers have found that the facial recognition system exhibits 99% accuracy when the person is a white man, while the model was only 35% accurate for dark-skinned women [147] and that this unfairness is stemmed from the lack of dark-skinned women in the model's training dataset. The issue is that "Big Data" does not necessarily provide fair and trustworthy models. For example, social media is a well-known source of large-scale dataset. However, it is reported that only 24% of online teens use Twitter [148]; therefore, conclusions extracted using the Twitter data are not necessarily correct.

It's not always the case that an unfair model performs worse on a population subgroup. The model might be accurate but still unfair. In this case, the dataset itself is biased, and the model replicates or amplifies the inherited bias. For example, Natural Language (NLP) models are often trained on a large body of human-written text (*e.g.*, article news). However, it has been found that word embeddings trained on Google News articles exhibit female/male gender stereotypes. For example, the researchers showed that the models answered that a father is a doctor as the mother is a nurse, or a "man" is a "computer programmer" while a "woman" is a "homemaker". This kind of bias occurs when a model is trained on data that is itself biased due to unfair systems or structures [149].

Yelp's restaurant review system is another example of AI bias. Yelp allows restaurants to pay to promote their establishments on the Yelp platform, but this naturally affects how many people see advertisements for a given restaurant and, thus, who chooses to eat there. In this way, Yelp reviews may be unfairly biased towards larger restaurants.

5.8.2 Automation bias

One of the renowned professors of psychology, Linda J. Skitka of the University of Illinois at Chicago, stated a clear definition of automation bias and that is "when many decisions are handled by automated aids (*e.g.*, computers, IoT devices, smart phones), and the human actor is mostly present to monitor on-going tasks, automation bias refers to a specific class of errors individuals tend to make in highly automated decision-making scenarios."

Automation bias is an over-reliance on automated aids and decision support systems. Automation bias was most likely a significant influence in the Enbridge pipeline catastrophe [150] on July 26, 2010, when massive volumes of crude oil were discharged into the Kalamazoo River and Talmadge Creek. The Enbridge oil pipeline catastrophe analysis reveals that complacency and automation bias played major roles. As a result, industry, policymakers, and regulators must consider automation bias while creating systems to limit the risk of complacent mistakes. It is the human tendency to take the road of least cognitive effort while leaning towards "automation bias". The same concept can be translated to the fundamental way that AI and automation work, which is mainly based on learning from large sets of data. This type of computation assumes that things will not be radically different in the future. Another aspect that should be considered is the risk of using a flawed training data then the learning will be flawed [151].

5.8.3 Other forms of (human-related) biases

5.8.3.1 Gender bias

Based on a survey report [152], the oil and gas industry is facing shortage of skilled worker while gender bias is making the situation worse by opting out female employees from recruitment. The report includes interviewing several male and female employees all around the world and analyzing their responses. In fact, the oil and gas is traditionally known as a male-dominated environment. However, some oil and gas companies are keen to ensure gender parity and workforce diversity, while others are leaving the gender gap to get wider. Although many companies seek to integrate gender parity in their policies, activities, and processes, they still fight a lot of issues like gender disparity and other forms of unconscious bias. The important observations of the mentioned study include following highlights:

- Women represent approximately one fifth of the employees in the oil and gas industry, a remarkably smaller share of the workforce than almost any other sector.
- The study found that the Oceania region, including Australia, was at the lower end for women being educated in the STEM (science, technology, engineering and mathematics) area, considered a critical entry point for technical level employment.
- While an average 22% of jobs in the industry are filled by women, a look at specific job categories tells a different story. University-educated

women hold fully 50% of entry-level office and business-support positions, but they hold only 15% of entry-level technical and field positions.

- The percentage of women in the industry's workforce drops over time and falls particularly sharply—from 25% to 17%—between the middle-management and senior-leadership career stages.
- 57% of women said that female employees receive less support for advancement into senior positions than the male employees; only 24% of men agreed. 56% of women said that women are overlooked for senior positions; only 23% of men agreed.
- Attracting and retaining greater numbers of women, particularly those with optimal backgrounds and skill sets will pose challenges for the industry.

According to this study, oil and gas companies are failing to fully leverage a critical pool of talent that could ultimately weigh heavily on the ability of the oil and gas companies to increase capital productivity.

Another study conducted by Australian trade companies [153] also highlights the opportunities local businesses are missing with a perceived lack of support for culturally diverse women. The study found that only 2% of directors in the trade sector are culturally diverse women. The report noted a number of key areas of concern including the general lack of access to sponsorship, in addition to lack of mentoring and networking for culturally diverse women. This is intensified by a mixture of gender and cultural biases that cause culturally-diverse women's lack of career progress and opportunities.

5.8.3.2 Cognitive bias

Cognitive biases [154] are a recently identified concept that is defined as mental flaws in our thinking and processing of information that may lead to incorrect or unreasonable judgments or conclusions. It was initially suggested in 1974 by Amos Tversky, and Daniel Kahneman in a Science Magazine article (Tversky and Kahneman, 1974 [155]). Numerous articles and research studies on cognitive biases and how they affect our perceptions and choices have been published since then.

An intuitive definition of cognitive bias is a subconscious error in thinking that leads to misinterpretation of information from the world and affects the rationality and accuracy of decisions and judgments. Biases are unconscious and automatic processes designed to make decision-making quicker and more efficient. Cognitive biases can happen by a number of reasons, such as social pressures and emotions. Over the last several years, there has been an increasing understanding of the dangers cognitive bias might pose to operational safety. For example, biases such as deviance, normalization, and group thinking are now generally acknowledged. In addition, the Deepwater Horizon [156] inquiry in 2010 raised broad awareness, at least within the offshore drilling sector, of the influence that cognitive bias may have on frontline thinking and decision making. The International Association of Oil and Gas Producers (IOGP) has raised the awareness about the critical nature of these cognitive difficulties for safety.

Cognitive bias is prevalent in the oil and gas industry. Consider a scenario where an exploration team is evaluating a potential location open to license bids. The data analysis focuses on a very effective analogy for describing the play for potential hydrocarbon well. Being optimistic about the play potential, the exploration team presents the opportunity to manage in very favorable terms that the company confidently bids for the construction of wells. Afterward, a new team reassessed the hydrocarbon potential after more data collecting and analysis. Given the ambiguity, the prospects and leads generated should have a broad range of uncertainty in petroleum production. Hence, cognitive bias leads to poor decision-making for the exploration team. The harmful effect of cognitive biases is not limited to the exploration phase. Project planning, development, production, and assessment initiatives face similar challenges.

5.9 Summary

The digital Industry 4.0 revolution has changed the landscape of the operational process of oil and gas immensely by utilizing advanced computing hardware and software technologies. The advancement, however, provides the opportunity for cyber criminals to become more efficient in finding loopholes in the IT, OT, or the intersection of these two. The heterogeneity and incompatibility of smart technologies and the connectivity problem between them creates another potential avenue for the cyber criminals. Both legacy and smart oil and gas industries face prominent challenges in adopting smart technologies, due to incompatibility of technologies purchased over time; lack of the technical familiarity and mindset in dealing with the smart systems; and issues related to human-to-machine and machine-to-machine interactions. Industry 4.0 has brought a major improvement to the oil and gas industry, however, industry leaders and practitioners must be aware that not everything about smartness of Industry 4.0 is bright! Blindly implementing or adopting smart technologies can potentially create numerous threats and vulnerabilities that we have encountered some of them in the past few years. It is imperative for the researchers and practitioners to consider these side-effects while embracing smart technologies.

CHAPTER 6

Designing a disaster management system for smart oil fields

6.1 Introduction and overview

6.1.1 Smart oil fields

Petroleum has been unarguably one of the essential elements of world economic growth throughout the past decades. For nearly two centuries, petroleum has been exploited as the primary natural source for many industrial products such as gasoline, natural gas, diesel, oil, asphalt, and plastic. Nonetheless, oil and gas (O&G) industries currently face several challenges mainly due to scarcity of petroleum reservoirs, requiring the companies to extract O&G at remote and adverse locations (*e.g.*, Golf of Mexico, Persian Gulf, West Africa) [157] where giant reservoir exist, hence, multiple oil extraction sites are built within a short distance. Operating at such remote sites is costly and constrained by limited crew and equipment. In addition, petroleum extraction is a fault-intolerant process and requires high-reliability *e.g.*, for drilling and downhole monitoring. Disasters, such as the Deepwater Horizon oil spill in 2010 [158], occurred due to faults in the safety system of the extraction. As such, strict regulations are being enforced by governments (e.g., U.S. environmental protection agency (EPA)) to prevent disasters and minimize the ecological impacts of O&G extraction.

To avoid potential flaws and disasters, oil fields are equipped with many cyber-physical devices (*e.g.*, sensors and actuators, as seen in Fig. 6.1), and the concept of *smart oil field* has emerged. Smart oil fields leverage numerous sensors, including those for temperature, Hydrogen Sulfide (H2S) gas emission, pipeline pressure, air pollution, and flow monitoring. These sensors gather a large volume of data (up to two Terabytes per day [159]) that most need to be analyzed and used in a real-time manner. Various research works [159–163] have collectively emphasized the need for smart oil fields with following requirements:

(1) Real-time decision-making during the extraction process to manage the drilling operation, which is challenging when the operation is controlled remotely by the management team.

IoT for Smart Operations in the Oil and Gas Industry
https://doi.org/10.1016/B978-0-32-391151-1.00015-0

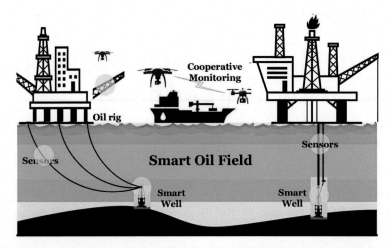

Figure 6.1 A smart oil field scenario including different oil extraction and monitoring sensors.

(2) Online monitoring of the site including rigs' structure, wells, and distribution lines to avoid any O&G leakage, corrosions identification, and future incidents prediction.

(3) Numerous sensors generate a large amount of data [164] that must be transferred to cloud datacenters for processing.

Services that utilize the sensors' data are categorized as either *latency tolerant* (a.k.a. nonurgent) or *latency intolerant* (a.k.a. urgent). Analyzing cost-efficiency of drilling, compressing and archiving captured surveillance videos, and generating weekly production reports are examples of nonurgent services. Such services are generally more compute intensive with less latency constraints. In contrast, pipeline pressure alarm and gas (*e.g.*, H2S) leakage detection are instances of latency intolerant services. Such services have real-time constraints and must be processed within a short latency (deadline). For instance, preserving workers' safety in an oil field requires processing the data generated by H2S sensors in less than five seconds [165].

6.1.2 Challenges of the current smart oil field solutions

Existing smart oil field solutions cannot meet the requirements of *remote* oil fields for the following two reasons:
1. Lack of reliable and fast communication infrastructure to onshore cloud datacenters for real-time processing of the extracted data.

2. Due to the harsh environment and shortage of manpower, more automation, and real–time processing is required to deal with abundant sensors and actuators.

Currently, remote smart oil fields use satellite communication to cloud and monitoring centers that are located on the mainland. However, satellite communication is known to be unstable and imposes a significant propagation delay leading to the latency in the order of seconds, which is intolerable for many real–time services in smart oil fields [166]. Current smart oil field solutions [167] do not consider the latency that exists in communication between the oil fields and cloud datacenters. As such, authors of this book studied the idea of smart oil fields in remote offshore sites. Their research proposes a smart solution using robust edge computing systems for remote smart oil fields with high latency connectivity to cloud datacenters. The edge computing system is defined as a micro data center that aims at handling latency–sensitive (*i.e.*, urgent) tasks. At the time of a disaster (*e.g.*, oil spill or gas leakage), different urgent activities must orchestrate to manage the disaster. For instance, real–time simulations must be conducted to predict oil spill expansion; emergency teams must be notified, and Unmanned Aerial Vehicles (UAVs) must be scheduled and dispatched for finer inspection [168]. However, edge computing resources are generally insufficient to handle such surges in demands.

6.1.3 Contributions of current research works

To overcome resource constraints of a single edge system and make it robust against the surge in demand, the smart solution leverages the edge computing systems available in nearby oil rigs, drillships, or even mobile micro datacenters and proposes a mechanism to engage them upon demand. Although federating edge systems can potentially mitigate the shortage of resources, such a federated environment involves new challenges that must be addressed to achieve the intended robustness. For a given task, the federated edge system imposes the stochastic transmission latency to a neighboring edge and the stochastic execution latency on the destination edge system. These latency times collectively are called *end-to-end latency*. For a latency–intolerant task, the end-to-end latency and its implied uncertainties must be captured so that the federated environment can be helpful. As such, the research problem considered in this study is *how to design a dynamic federated edge computing system that is robust against uncertainties that exist in both communication and computation and can handle surges in demand for latency-intolerant tasks during a disaster?*

To solve this issue, each edge computing system has a load balancer that maintains the federation view. The load balancer uses a probabilistic model to capture both computation and communication uncertainty in a federated context. The probabilistic model predicts the likelihood of fulfilling the arriving task's latency restriction. On the other hand, a probabilistic approach allows the load balancer to utilize the federation, making the edge system resilient to spikes in task arrival. Because the edge computing system is critical to the smart oil field, its robustness leads to a cleaner, more cost-effective O&G business.

In summary, The **contributions** of this research are as follows:

- Proposing a resource allocation model that dynamically federates edge computing systems to enable robust smart oil fields.
- Establishing a probabilistic model to capture end-to-end uncertainties exist in the federated edge environment and calculate the probability of success for tasks in this environment.
- Developing a federation- and QoS-aware resource allocation heuristic based on the probabilistic model.
- Analyzing the performance of the federated edge computing system under various workload conditions.

6.2 System model for IoT- and edge-based task allocation

The system model for IoT- and edge-based task allocation in a smart oil field, shown in Fig. 6.2, includes two-tier of computing systems where edge systems are located in local or in the first tier, and cloud datacenters are in the second tier. An *edge system* is defined as a set of machines with limited computing power, storage, and communication capacity [169] that work together under the same management platform (*i.e.*, resource manager) to offer various services required at the oil field. In analogy with gigantic cloud datacenters, the edge systems are known as micro/mini datacenters [170]. The edge system can be located within an oil rig structure or mounted on a drill-ship near the rig [171]. The edge systems are protected with temperature and water-resistive materials in disaster-prone environments. As such, the typical assumption is that the edge system itself is safe from oil field disasters.

Each edge system has a load balancer module that allows tasks to be sent to the suitable execution platform (*e.g.*, central cloud, edge system). This smart solution also proposes using a wireless network to connect the load balancer to its peer edge systems. The load balancer decides whether

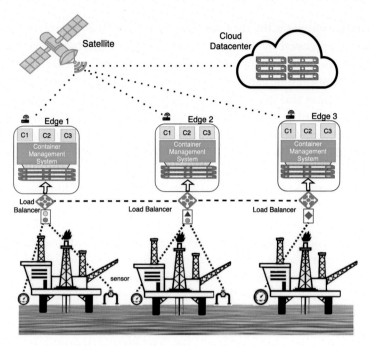

Figure 6.2 Architecture of federated edge computing of smart oil field where sensor-generated tasks are sent to edge nodes for execution.

to assign (offload) arriving tasks to peers or the central cloud based on workload intensity. Nonurgent work should be processed in cloud data centers, while urgent tasks should be processed at the edge. Urgent work should go to edge federation, nonurgent to cloud datacenters. This study considers edge system heterogeneity. That means certain edge systems are faster than others (*i.e.*, processor cores and available memory).

Different types of sensors (*e.g.*, temperature sensor, pressure sensor, gas sensor, camera) generate data that are consumed by heterogeneous tasks (defined as *task types*) to offer various services required in a smart oil field. The task types (*e.g.*, image processing for oil spill detection, toxic gas detection, weekly report generation) are assumed to be limited and known in advance. Also, each task is assumed to be independent of other tasks. The format [172] and size of generated data by some sensors can potentially vary, whereas, for some other sensors, they are constant. For example, images captured by cameras to detect oil spill can be of different sizes. This randomness serves as one primary reason for uncertainty in the execution time of the task type that processes images to detect an oil spill [173].

A contrary example is the data periodically generated by temperature sensors and is processed by a task type that identifies fire hazards in the oil field [174]. In the latter example, even though the data size does not vary, the task execution time can have uncertainty due to the workload of neighboring machines and multi-tenancy [175]. Apart from the execution time uncertainties exist within each edge system, due to heterogeneity, a task of a specific type can have different (*i.e.*, uncertain) execution times across different edge systems.

When a task of type i arrives at an edge system j, an individual deadline is assigned (denoted δ_i) based on its arrival time and maximum latency tolerance. Notably, task deadlines vary depending on the service type offered. Furthermore, tasks arrive at an edge system in random order. This study focuses on surge demands in the edge system that overload (oversubscribe) it. Thus, the task arrival rate to the edge system is so high that meeting all deadlines is impossible. The assumption is that each task arrives sequentially and requires only one processing unit (e.g., a processing core). To maximize the robustness of the edge system, a resource allocation method aims to maximize the number of tasks meeting their deadline constraints.

6.3 Robust resource allocation using federation of edge computing systems in remote smart oil fields

The synopsis of the proposed resource allocation model in the federated edge computing system is demonstrated in Fig. 6.3. A load balancer module is the primary enabler of edge federation in the resource allocation architecture. Every edge system contains a load balancer that identifies the best edge system (either the receiving edge or an adjacent one) for each arriving task. The functionality of the load balancer is particularly prominent to cope with the uncertainty that exists in task arrivals (*e.g.*, during disaster time) and make the edge system robust against it. The load balancer operates in *immediate mode* [176] and assigns arriving tasks to the appropriate edge system immediately upon task arrival. The appropriateness is characterized based on the edge system that maximizes the probability of the task meeting its deadline (known as the *probability of success*). The probability of success for task t_i with deadline δ_i can be calculated for each neighboring edge system by leveraging the end-to-end latency distribution of executing task t_i on that system. To avoid repetitive task reassignment and compound latency, this study proposed that the task cannot be re-allocated once a task assignment is made.

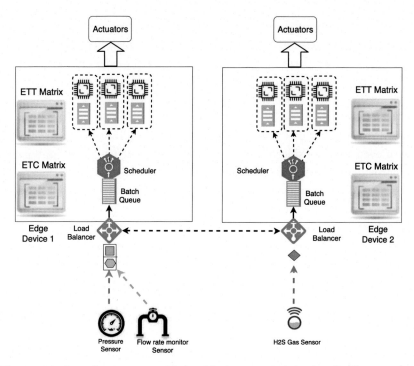

Figure 6.3 An edge system with load balancer module that facilitates edge federation. Task requests generated by sensors are received by the load balancer module and are assigned to the edge system that maximizes the likelihood of success for the task.

The resource allocation of each edge system leverages the historical information of computational and communication latencies to build Probability Density Function (PDF) of their distributions. For that purpose, each load balancer maintains two matrices, namely Estimated Task Completion (ETC) [177] and Estimated Task Transfer (ETT), to keep track of computational and communication latencies for each task type on each neighboring edge system. Entry $ETC(i,j)$ keeps the PDF of computational latency for task type i on edge system j. Similarly, entry $ETT(i,j)$ keeps the PDF of communication latency for task type i to reach edge system j. The entries of ETC and ETT matrices are periodically updated in an offline manner and they do not interfere with the real-time operation of the load balancer.

Upon arriving task t_i, load balancer of the receiving edge can calculate the end-to-end latency distribution of t_i on any neighboring edge j, using $ETC(i,j)$ and $ETT(i,j)$. The end-to-end distribution can be used to obtain the probability of completing t_i before its deadline, denoted $p_j(t_i)$, on any of those edge systems. The formal equation: $p_j(t_i) = \mathbb{P}(E_i \leq \delta_i)$. Hence, the probability calculation for task t_i on the receiving edge does not imply further communication latency. As such, for the receiving edge r, the probability: $p_r(t_i) = \mathbb{P}(M_i \leq \delta_i)$. In the next step, the edge system that provides the highest probability of success is chosen as a suitable destination to assign task t_i. This implies that task t_i is assigned to a neighboring edge system, only if even after considering the communication latency, the neighboring edge provides a higher probability of success. It is noteworthy that the probability of success on a neighboring edge can be higher than the receiving edge by a nonsignificant amount. In practice, a task should be assigned to a neighboring edge, only if the neighboring edge system offers a substantially higher probability of success. To understand if the difference between the probabilities is substantial, this work leverages confidence intervals (CI) of the underlying end-to-end distributions, from which the probabilities of success for receiving and remote edges are calculated. More specifically, this research determines a neighboring edge offers a significantly higher probability of success for a given task, only if CI of end-to-end distribution of the neighboring edge does not overlap with the CI of end-to-end distribution of the receiving edge.

6.4 Performance evaluation of resource allocation method

6.4.1 Experimental setup

This research used EdgeCloudSim [178], which is a discrete event simulator for performance evaluation. The researchers simulate five edge systems (micro-datacenters) each one with eight cores and [1500, 2500] Million Instructions Per seconds (MIPs) computing capacity. Cores of each edge system are homogeneous: however, different edge systems have different MIPs that represents the heterogeneity across the edge systems. Additionally it considers a cloud datacenter with 40,000 MIPs to process nonurgent tasks. Task within each edge is mapped in the first come first serve manner. The bandwidth to access cloud is based on satellite communication and set to 200 Mbps, and the propagation delay is 0.57 seconds [173]. In each workload trial generated to simulate the load of a smart oil field, half of the tasks represent the urgent category, and the other half represent nonurgent tasks. Each task is of a certain type that represents its service type. In each workload trial, urgent tasks are instantiated from two different task types, and nonurgent tasks are instantiated from two other task types. The execution time of each task instantiated from a certain type is sampled from a normal distribution, representing that particular task type. Each task is considered to be sequential (requires one core), and its execution time is simulated in the form of MIPs. Poisson distribution (with different means for different task types) is used to generate the inter-arrival rate of the tasks and simulate task arrival during oversubscription periods. The number of tasks in each workload trial is varied to represent different oversubscription levels.

Deadline for task i in a workload trial is generated as: $\delta_i = arr_i + \beta \cdot avg^i_{comp} + \alpha \cdot avg^i_{comm} + \epsilon$, where arr_i is the task arrival time, avg^i_{comp} is average computational latency of the task type across edge systems, and avg^i_{comm} is average communication latency. β and α are coefficients, respectively, represent computation and communication uncertainties, and ϵ is the slack of other uncertainties exist in the system. This research proposes maintaining ETC and ETT matrices in every edge system and updates them in every 10% of the workload execution. The entries of these matrices are considered as normal distribution as mentioned in the system model. For accuracy, each experiment was conducted 30 times and the mean and 95% confidence interval of the results are reported.

6.4.2 Baseline task assignment heuristics for load balancer

Minimum Expected Completion Time (MECT): This heuristic [179] uses the ETC matrix to calculate the average expected completion time for the arriving task on each edge system and selects the edge system with the minimum expected completion time.

Maximum Computation Certainty (MCC): This heuristic (used in [180]) utilizes ETC matrix to calculate the difference between the task's deadline and average completion time (called certainty). Then, the task is assigned to the edge that offers the highest certainty.

Edge Cloud (EC): This heuristic operates based on conventional edge computing model where no federation is recognized. Specifically, urgent tasks are assigned to the receiving edge and nonurgent tasks are assigned to the cloud datacenter.

6.4.3 Experimental results

6.4.3.1 Analyzing the impact of oversubscription

The main metric to measure the robustness of an oversubscribed edge system in a smart oil field is the deadline miss rate of tasks. This experiment

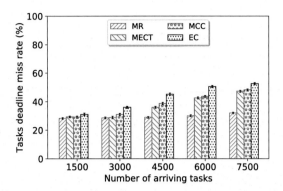

Figure 6.4 The impact of increasing oversubscription level (number of arriving tasks) on deadline miss rate using different task assignment heuristics in the load balancer.

studies the performance of the proposed system by increasing the number of tasks sensors generate (*i.e.*, oversubscription level). Fig. 6.4 shows the results of varying the number of arriving tasks (from 1,500 to 7,500 in the horizontal axis) on deadline miss rate (vertical axis) when different task assignment heuristics are applied.

In Fig. 6.4, it is visible that as the number of tasks increases, the deadline miss rate grows for all of the heuristics. Under low oversubscription level (1,500 tasks), Maximum Robustness (MR-proposed heuristic), MECT, and MCC perform similarly. However, as the system gets more oversubscribed (4,500 tasks) the difference becomes substantial. With 7,500 tasks, MR offers around 16% lower deadline miss rate than MECT and MCC and approximately 21% better than EC. The reason is that MR captures end-to-end latency and proactively utilizes federation, only if it has a remarkable impact on the probability of success. Nonetheless, EC does not consider federation, and other baseline heuristics only consider the computational latency. Therefore this experiment concludes that considering end-to-end latency and capturing its underlying uncertainties can remarkably improve the robustness, particularly, when the system is oversubscribed (*e.g.*, at a disaster time).

6.4.3.2 Analyzing the impact of urgent tasks ratio

In this experiment, while the system is oversubscribed with 8,000 tasks, the percentage of urgent tasks in the workload is varied from 10% to 90% (horizontal axis in Fig. 6.5) and in each case the deadline miss rate (verti-

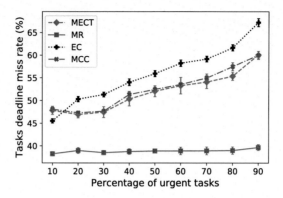

Figure 6.5 Impact of increasing urgent tasks on deadline miss rate.

cal axis in Fig. 6.5) is measured. According to Fig. 6.5, EC provides lower deadline miss rate than MECT and MCC at 10% the urgent tasks (*i.e.*, 90% nonurgent tasks). The reason is that EC redirects nonurgent tasks to the cloud datacenter and the remaining urgent tasks can complete on-time on the receiving edge system. Although MECT and MCC use the cloud for nonurgent tasks too, they utilize federation for some of the remaining urgent tasks without considering end-to-end latency. Hence, their performance is degraded. However, for more than 20% urgent tasks, EC performs worse than MECT and MCC, because it cannot utilize the federation. In all cases, it is observed that MR outperforms other heuristics due to consideration of end-to-end latency and its underlying uncertainties.

6.4.3.3 Analyzing communication overhead of edge federation

Although the result of previous experiment state that using federation improves system robustness, the communication overhead of task assignment in the federated environment is uncertain. Therefore this experiment evaluates the communication latency imposed as a result of applying different task assignment heuristics. Specifically, this experiment measures the mean communication latency overhead (vertical axis in Fig. 6.6) induced to each task, for the various numbers of arriving tasks.

Fig. 6.6 shows that MECT and MCC cause higher average communication latency. The reason is that these heuristics do not consider the communication latency and aggressively redirect tasks to the same edge system, making that particular network link (between receiving edge and redirected edge system) congested. In contrast, MR that considers communication la-

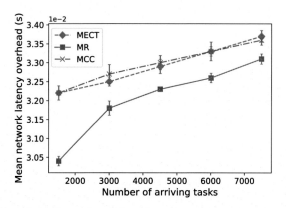

Figure 6.6 Mean communication latency overhead introduced to each task in edge federation by different heuristics.

tency and redirect tasks more conservatively, only if the improvement in the probability of success is substantial.

6.4.3.4 Analyzing average makespan of tasks

Different task assignment heuristics cause various computational latencies for the tasks. To understand the computational latency, this experiment measures the average makespan of tasks, resulted by applying various task assignment heuristics.

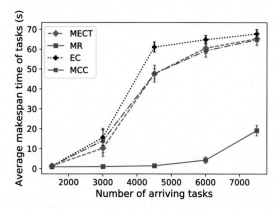

Figure 6.7 Average makespan of tasks using various task assignment heuristics.

Fig. 6.7 demonstrates that EC leads to the maximum average makespan time. The reason is that EC does not utilize federation, making the receiving edge system highly oversubscribed while other neighboring edge systems are underutilized. Hence, average makespan time rapidly rises after the receiving edge is saturated with 3,000 tasks. MECT and MCC do not consider the stochastic nature of task completion time; hence, they can potentially assign arriving tasks to one edge and oversubscribe that. As a result, the average makespan of tasks rises. In contrast, MR considers stochastic nature of end-to-end latency and calculates the probability of success on neighboring edge systems. Besides, it assigns tasks to a neighboring edge system, only if it offers a sufficiently higher probability of success. Hence, MR offers the lowest average makespan time than other heuristics.

6.5 Summary

In this research work, the goal was to provide a smart oil field that is robust against disasters and surges in real-time service requests. To achieve that, the proposed solution presented dynamic federation of edge computing systems, exist in nearby oil fields. Within the federated environment, two sources of uncertainty, namely communication and computation, were captured that are otherwise detrimental to the real-time services. The federation is achieved by a load-balancer module in each edge system that is aware of the end-to-end latency between edge systems and can capture the stochasticity in it. The load balancer leverages this awareness to find the edge system that can substantially improve the probability of success for each arriving task. Experimental results demonstrate that proposed federated system can enhance the robustness of edge computing systems against uncertainties in arrival time, communication, and computational latencies. The smart load balancer could be particularly useful (by up to 27%) for higher levels of oversubscription. Even for naïve load balancing methods (MCC and MECT) in the federation, the performance improvement is approximately 13%.

CHAPTER 7

Case study I: Analysis of oil spill detection using deep neural networks

7.1 Introduction

The ecosystem of the coastal area is significantly hampered by the oil spill occurrence, which is one of the primary concerns for environmental pollution. Due to different incidents (*i.e.*, malfunction in oil extraction, broken pipelines, deliberate discharge of tank–cleaning wastage) in oil fields or other oil transportation system carriers (*i.e.*, ships, boats, drilling ship) starts to impact the environment gradually. This event can lead to disasters like Deepwater Horizon [158] as such oil spills have significant adverse impacts on the environment as well as the economic backbone of coastal areas. Therefore, identifying oil spills accurately and starting recovery steps as early as possible play a vital role in the contingency plan of the oil spill prevention system. With remote sensing technology [181] and synthetic aperture radars (SARs), capturing images of remote sea areas has become easier than before. In SAR images, black or dark spots are captured as oil spills in sea areas. Although, these dark areas can fall into a similar category of oil spill rather than original oil spills, which creates the oil spill identification problem challengingly complex. This similar but not original oil spill category in SAR images is considered as look-alike [182].

The oil spill identification problem is tackled in various ways in the literature. Major two approaches are classification [183] and Semantic segmentation [184]. Although the classification of oil spills from SAR images has been suggested in the literature, semantic segmentation can provide a more accurate identification of oil spills, improving the recovery system. As semantic segmentation refers to the pixel-wise classification of the input images, it can be used to train neural networks to identify each pixel of an image to a particular class. Especially when one has a dataset with pixel-wise ground truth labeling, semantic segmentation can be one of the potential solutions. Moreover, utilizing the deep neural network in oil spill detection and evaluating the performance of the models are also challenging due to proper datasets and benchmarking. Most of the previous research

IoT for Smart Operations in the Oil and Gas Industry
https://doi.org/10.1016/B978-0-32-391151-1.00016-2
177

has been conducted using a custom dataset and with heavy preprocessing [185]. Therefore, comparing different deep neural network (DNN) models' performance becomes challenging for researchers even if many states of the art DNN–based semantic segmentation architectures evolved with greater accuracy for different popular datasets.

To overcome this challenge, a research group from MKLAB has published a well-defined and recent oil spill dataset [186] that can be utilized for analyzing different popular DNN models. Using this dataset, researchers have already performed various experiments using different popular DNN architectures (*i.e.*, UNET, LINKNET, PSPNET). Being a relatively new SAR dataset for oil spill detection, the proposed dataset is not yet tested with prevalent models of DNN like FCN-8s. Although the dataset is preprocessed with actual ground truth, other preprocessing (*i.e.*, resizing, identifying labels) needs to be done to feed into desired DNN models. This research investigates the performance of the FCN-8s model on this very new oil spill dataset. Hence, this work aims to implement an FCN network for semantic segmentation that is not tested before with the recently published oil spill dataset. As such, the contribution of this case study can be listed in the below sentences.

1. Dataset preprocessing (*i.e.*, label identification, image resizing) for oil spill detection FCN-8s model.
2. Implement and analyze FCN-8s architecture for multi-class semantic segmentation with oil spill dataset.
3. Thorough analysis of experimental results with respect to dataset and different optimizers.

7.2 Data acquisition

The lack of a common dataset for oil spill detection is an obstacle for the research community investigating and comparing appropriate DNN network models. Previous researchers [15,16,27] limited their activities to customized datasets adapted based on their related analysis. However, because each method uses a different dataset, the presented results are not comparable. As a result, there is no common comparison point. Given this shortcoming, this work utilizes one of the well–organized and semantically annotated satellite image datasets [186] for the comparative analysis. The data acquisition for oil spill detection mainly comprises capturing images via synthetic aperture radar (SAR) technology using satellites and performing image processing on them. In a nutshell, satellite SAR images of oil-polluted sea areas were collected using the Copernicus Open Access Hub database of the European Space Agency (ESA). In addition, the European Maritime Safety Agency (EMSA) provided information about the pollution event's geographic coordinates and timestamps via the CleanSeaNet service. As a result, the EMSA records confirm that the dark spots depicted in the SAR images are oil spills, resulting in a solid ground truth subset. The oil pollution data is from September 28, 2015, to October 31, 2017, and the SAR images are from the European Satellite missions Sentinel-1 and Sentinel-2.

Sentinel-1 satellites are installed with a C-band SAR system. With pixel spacing of 10*10 meters, the ground range coverage of the SAR sensor is approximately 250 kilometers. These specifications show the SAR sensor's ability to cover large areas of interest while also indicating the difficulty of capturing small objects, such as ships. The radar image's polarization is dual, with vertical polarization transmitted—vertical polarization received (VV) and vertical polarization transmitted—horizontal polarization received (VH). Only the gathered raw data from the VV band was processed to create the SAR image dataset, with a series of preprocessing steps to extract standard visualizations. The following phases were included in the preprocessing plan:

- According to EMSA records, every confirmed oil spill was located accurately.
- A region containing oil spills and possibly other contextual information of interest was cropped from the raw SAR image. The cropped image was rescaled to 1250*650 pixels in resolution.
- Radiometric calibration was used to project each of the 1250*650 images into the same plane.
- The sensor noise scattered throughout the image was suppressed using a speckle filter. Because speckle noise has a granular texture, it was suppressed using a 77% median filter.
- For converting the actual luminosity values, a linear transformation was used.

7.2.1 SAR images

Remote monitoring systems for sea areas have significantly evolved with synthetic aperture radar (SAR) technology. SAR sensors are usually mounted on aircraft or satellites that can generate an image covering a wide range of areas of earth or sea. Due to its all-weather stability and workability in different illumination conditions, SAR images have become one of the most reliable sources of oil spill detection. The SAR sensor sends radio wave pulses to the target area, and the reflections are received to generate the image of the target area. As oil spills spread over the sea surface by reducing the sea's currents, the reflected radio pulses create dark spots in the SAR images. As such, the vivid region of the images represents the unspoiled areas of the sea. Although dark or black spots represent the possibility of oil spills, it does not always correspond to the ground truth. Dark spots can also represent regions with meager wind speed, weed beds, algae blooms, wave shadows behind land, etc. [187,188]. With similar dark spots, which is considered a look-alike in literature, the oil spill detection problem poses one extra layer of complexity.

7.3 Dataset overview

A total of 1112 images were extracted from the raw SAR data as part of this process. Oil spills, look-alikes, ships, land, and sea surface are all represented in the images, with the latter always considered a background class. Each image is annotated using information from the EMSA records and human identification for maximum efficiency. Because the presented dataset was designed to be used with semantic segmentation methods, each of the five classes was assigned a different RGB color. As a result, the ground truth masks that accompany the images of the dataset are colored according to the identified class for each instance of interest.

The annotated dataset was divided into training (90%) and testing (10%), with 1002 and 110 photos, respectively. There is a substantial imbalance between the occurrences of the different classes due to the nature of the oil spill detection problem over SAR pictures. More particular, sea surface or land class samples are predicted to dominate the dataset. Oil spills and look-alikes, on the other hand, are generally extended in smaller regions of SAR photos. Due to natural phenomena such as low wind speeds and wave shadows close to land, look-alikes can span a larger region. Because their presence in an oil pollution scenario is not assured, the "ship" class samples are projected to enumerate much fewer instances.

7.3.1 Challenges of oil spill datasets

One of the vital challenges of comparing different DNN architectures for oil spill detection is the lack of well-defined and specific datasets for the oil spill. Many researchers have performed various deep neural network techniques to detect oil spills. Although, they used their customized dataset for a specific solution, which creates the problem of comparing different solutions for a generic dataset. To overcome this problem, this research work has used one of the recent datasets, which has been published by one of the research groups of Greece named MKLab [186]. To the best of our knowledge, there is no such oil spill dataset preprocessed well enough with ground truth pixel-wise labeling like MKLab's oil spill dataset. Hence, this case study chooses this dataset to use oil spill detection techniques utilizing DNN methods that were not tested earlier.

7.4 Machine learning models

There have been various classification methods suggested by literature for Oil Spill detection with SAR images. The majority of them involved three primary steps [186]. They are (a) Dark spot detection from SAR image, (b) Feature extraction from identified portion of an image, and (c) Oil spill/look-alike classification. In the above, binary segmentation is utilized in the first step to identify black spots. Then statistical features are extracted in the second phase from the identified segment. Finally, the considered region or whole image is classified as an oil spill/look-alike. Considering the above-mentioned three-step method for classification between the oil spill and look-alike, an automated framework has been introduced by Solberg et al. in [189]. Furthermore, authors in [190] presented a probabilistic method where they utilize statistical data received from earlier measurements of physical and geometrical attributes of the oil spill and look-alike for evaluating a sample image to develop a template for detection. Then, using this template classification algorithm is performed with two different techniques to compare their results. Finally, one of the interesting research has been conducted by authors in [191], where they include wind speed data to enhance the classification method with the approximation of oil spill period of occurrence.

Other than classification, semantic segmentation with DNN methods has been widely applied in oil spill detection. Especially the very first utilization of a neural network for identifying relevant labels from SAR images has been introduced by Fabio et al. in [192]. For improving the internal characterization of the input, a neural network is used in the feature extraction process by De Souza et al. in [193] where different algorithms of speckle filtering are studied for image segmentation. Konstantinidou et al. in [194] proposed Orfeo Toolbox SVM vector classifier for the oil spill and look-alike classification, where segmentation is utilized for feature extraction, and extracted features are used in the classification process. Different shallow-level descriptors and filters are used to bring out texture information from dark spots. The extracted data is applied to a texture-classifying neural network algorithm to identify spatial characteristics of oil spills in [195].

Most of the works mentioned earlier provided solutions considering oil spill detection as a binary classification problem. The input image is entirely labeled, or part of it is identified as an oil spill or look-alike. Hence, much important information (*i.e.*, oil discharging ship, lands near oil spill polluted area) captured in SAR images was not utilized to help improve the detection system or warning systems for oil spills. SAR images can capture a wide range of areas where many important classes include a multi-class semantic segmentation solution that is more appropriate than a binary classifier. Consequently, authors in [196] proposed the DNN method to produce a segmentation label and suggested an f-divergence function as a means of evaluating segmentation results with ground truth. This function is used while generating segmentation masks from the input. Thus, the authors tried to minimize the f-divergence to improve the segmentation results. Although, the proposed method considered only oil spill class segmentation rather than per-pixel classification for multi-class problems. One of the research work considering multi-class segmentation proposed by Gallego et al. in [197], where the Side-Looking Airborne Radar (SLAR) images were utilized. In this work, the main limitation was the dataset with fewer instances for training a DNN model. Therefore, fewer research works are available in the literature for oil spill detection problems with a multi-class segmentation approach. Additionally, due to not having a well-defined and preprocessed dataset for oil spill detection with DNN models, researchers had to preprocess and customize different datasets to feed into their own models. On the other hand, different evaluation techniques have been utilized by researchers, which creates the problem uncomparable concerning DNN models and evaluation criteria.

7.4.1 Semantic segmentation

According to Wikipedia, image segmentation partitions an image into multiple segments or pixels. Furthermore, semantic segmentation is a process of understanding from coarse to acceptable inference where the main idea is originated from classification. It is the process of identifying each pixel of an image that belongs to a particular class. This can be considered the task of relating each pixel in an input image to a class label where these labels can be any object like a car, flower, furniture, or even a person. Semantic segmentation can be beneficial with the data set of arbitrarily shaped classes (*i.e.*, classes that do not have any particular geometric shape) that can only be identified from pixel-wise labeling.

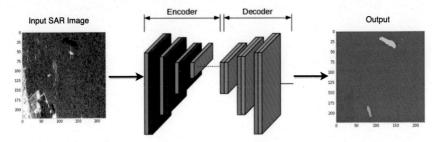

Figure 7.1 Encoder decoder architecture for semantic segmentation of SAR image.

Among various applications of semantic segmentation, self-driving cars, robotic systems, and damage detection are popular fields. In neural networks, there have been many implementations of model architectures (*e.g.*, Fully Convolutional Network (FCN), U-Net, Mask RCNN, Segnet) to perform semantic segmentation. In general, the majority of the architectures follow the encoder–decoder network, as shown in Fig. 7.1 to perform semantic segmentation. Typically for the encoder, any pretrained classification network (*e.g.*, VGG, ResNet) can be utilized to extract the features for classification. On the contrary, the decoder semantically extrapolates the distinguishable features comprehended by the encoder onto the pixel space for pixel-wise classification of the original image.

7.4.2 Oil spill detection DNN model overview

This case study considers oil spill detection as a semantic segmentation problem of pixel-wise classification from SAR images (*i.e.*, MKlab's dataset). As the dataset has labeling of ground truth, identifying each pixel belonging to the identical class is the main problem of this study. Furthermore, having the class labels in the dataset enables the idea of implementing a supervised learning model in the solution domain. Considering the computational perspective of the deep neural network models, Fully Convolutional Networks (FCN) are one of the potential candidates for semantic segmentation [198]. There are three versions of FCN networks, namely FCN8, FCN16, and FCN32. The main differences among these architectures are the stride of the final convolution layers of the decoder and skip links for the upsampling resolution of the input image. Although, for the encoder part, the three FCN models utilize the same architecture.

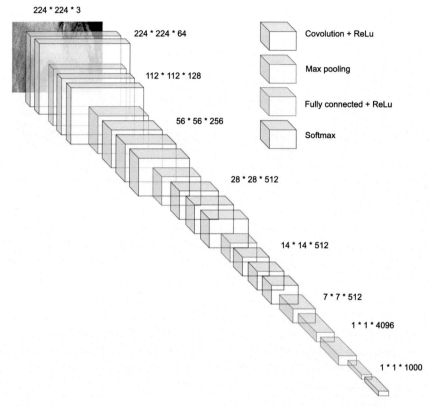

Figure 7.2 VGG16 architecture for image classification that reduces the input size of the image and loses spatial information in deep layer of convolution [199].

As depicted in Fig. 7.2 with the development of efficient classification models such as Visual Geometry Group (VGG) [200], the DNN classification sector has improved immensely. VGG creates a pile of convolution layers, including down–sample layers (*i.e.*, max-pooling). The last step stacks fully connected layers, which allow the network to learn global information. However, it loses the localized information of the global data that is not relevant to the classification problem. In contrast, it needs to identify or extract localized information from the global data in semantic segmentation. Finding the global data's local information is considered a decoder in the semantic segmentation problem. Thus, the spatial information needs to be utilized in the pixel-wise classification for segmentation work.

7.4.3 Oil spill detection DNN model architecture

The primary motivation of this work is to use the fully convolutional network for semantic segmentation [198]. In fully convolutional networks, there are no fully connected layers in the network, which reduces the number of parameters and computation time. That leads to a less computational expensive DNN network. Furthermore, having all local connections, the network can perform its operation even with arbitrary input image size. As mentioned earlier, segmentation architecture typically has two parts to get the pixel-wise mapping of given classes. They can be defined as:

1. Downsampling path (Encoder): This part extracts the semantic information from the input image.
2. Upsampling path (Decoder): This part retrieves the spatial information from semantic information of encoder output.

7.4.3.1 Encoder

Based on the concept of the unrolling back of the VGG net for pixel-wise classification, this research is motivated to utilize the VGG net as the decoder. Thus, the DNN model used in this work is FCN-8s [198], where VGG16 net is used to build the encoder part. As Fig. 7.3 demonstrates, the encoder part of the model has five blocks where each block represents convolution and pooling. These encoder layers downsample the image in every block to half of its previous layer. Therefore, in five blocks of the encoder, the input image is reduced to a lower resolution where feature and context (*i.e.*, what) extraction occurs, although the spatial information (*i.e.*, where) of the context for the input image is lost in this process implicitly.

7.4.3.2 Decoder

In the decoder part, the image has to be upsampled from low to high resolution, and each pixel is labeled with the corresponding class. Different upsampling methods (*i.e.*, nearest neighbor, interpolation, Gaussian reconstruction) can be utilized for the decoder part. These are different processes to connect rough outputs to dense pixels. An example presented in [198], bilinear interpolation estimates an output y_{ij} from its closest four inputs by a linear map that relies on the respective positions of the input and output

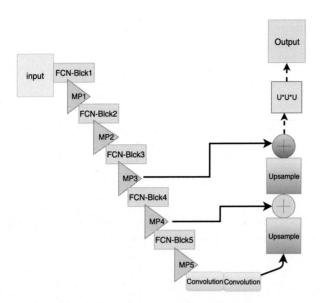

Figure 7.3 FCN-8s architecture for semantic segmentation where the first five blocks work as encoder and decoding is done with three upsampling blocks.

cells. Therefore, y_{ij} can be defined in the following equation:

$$y_{ij} = \sum_{\alpha,\beta=0}^{1} \left| 1 - \alpha - \frac{i}{j} \right| \left| 1 - \beta - \frac{i}{j} \right| x_{\lfloor \frac{i}{j} \rfloor + \alpha, \lfloor \frac{i}{j} \rfloor + \beta}, \qquad (7.1)$$

In the above equation, f represents the upsampling factor. The main idea is that upsampling with factor f represents convolution with a fractional input stride of $1/f$. Considering f as the integral, it's typical to devise upsampling along "backward convolution" by contrasting the forward and backward passes of more particular input-strided convolution. Hence, for end-to-end learning, upsampling is performed in the decoder by backpropagation from the pixel-wise loss. In FCN-8s, backward convolution is exploited with some output stride for upsampling. As presented in Fig. 7.4, this convolution, also known as deconvolution or convolution transpose, where the output size gets larger, reflecting the opposite direction of convolution. Hence, FCN–8s utilizes an upsampling path to understand the precise localization (*i.e.*, where the information is located in the image) known as a decoder in semantic segmentation.

As demonstrated in Fig. 7.3, the FCN–8s model upsamples (upsample block) the lower resolution image to higher resolution with convolution

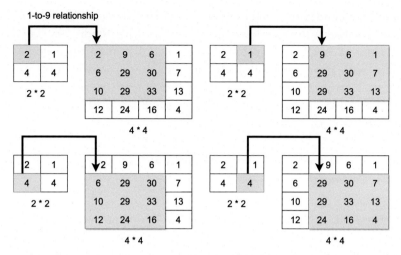

Figure 7.4 Deconvolution explained by Naoki Shibuya in medium [201] using upsampling process. Here, a 2 × 2 matrix is upsampled to 4 × 4 matrix maintaining 1-to-9 relationship.

transpose. These blocks of upsampling are sometimes considered as "unpooling" that is parameter-free. Additionally, utilizing deep learning, these "unpooling" blocks can be trained with powerful functions from the substantial dataset. The FCN-Xs model proposed paper considers these convolution transpose or deconvolution layers to have a trainable kernel. The last upsampling step enlarges the resolution by $2^3 = 8$; that is why the model is named FCN-8s. From the architectural point of view, FCN-8s remove the final classification layers of the VGGnet and add fully convolutional layers as a decoder in the architecture. While decoding the input image to its original form, the model learns the pixel-wise mapping used in the inference part.

7.5 Performance evaluation of oil spill detection

7.5.1 Experimental setup

This study used the Keras [202] framework to implement the FCN-8s model. Being a high-level neural network API, where the building block is composed of python, Keras can execute its operation on top of Tensorflow, CNTK, or Theano. Moreover, being user-friendly, modularity, and extensible, Keras can perform fast experimentation with DNN models. For training and evaluation of the FCN-8s model, Tesla k40c GPU with 12 GB memory is used as the model has memory issues with Nvidia 1050Ti GPU (*i.e.*, memory 4 GB). With 200 epochs, it took around three hours for training with a batch size of 32. While using 100 epochs with batch size 16, it took about 2 hours to complete the training phase, although the result of the model was not good enough, which is why 200 epochs were chosen for training the FCN-8s model.

7.5.2 Preprocessing dataset for FCN-8s

7.5.2.1 Original dataset

The oil spill dataset of MKlab consists of 1112 images that were preprocessed and extracted from raw SAR data. In the preprocessing step, every verified oil spill was labeled as ground truth according to EMSA records. Next, verified images were cropped and rescaled to 1250*650 pixels, and radiometric calibration was applied to project every image into the same plane. Finally, a speckle filter was used to deal with the sensor noise, and the linear transformation was applied to convert luminosity values from dB to real.

Table 7.1 Class imbalance in oil spill dataset.

Class	Pixels
Sea Surface	797.7M
Oil Spill	9.1M
Look–alike	50.4M
Ship	0.3M
Land	45.7M

The images in the dataset have instances belonging to 5 classes, including sea surface, oil spill, look–alike, ship, and land, where the sea surface is considered the background. The combined information (EMSA records & human verification) is used for image annotation for greater efficiency. In the case of training and evaluation, pixel–wise labeling of identified classes is provided as 1D labels. The dataset is split into training and testing with a ratio of 90%:10%, which is 1002 and 110 images, respectively. One of the major challenges of oil spill detection from SAR images is the class imbalance problem, where sea surface and land class have a significant impact. This problem of class imbalance is reported in the dataset with pixel counts from each individual class in Table 7.1.

7.5.2.2 Identify and labeling classes in dataset

From the 1D label of the dataset, each class was identified and then labeled the pixel of the particular class with five different colors that are represented in Table 7.2. The five classes of the dataset are appropriately identified from its 1D labels and reported in Fig. 7.5. The FCN–8s model is implemented to have an input size of 224 × 224 resolution images, and the images were resized from their original size (1250 × 650) to 224 × 224 that is presented in Fig. 7.6.

Table 7.2 Pixel-wise class coloring.

Class	Color
Sea Surface	Red
Oil Spill	Yellow
Look–alike	Green
Ship	Blue
Land	Purple

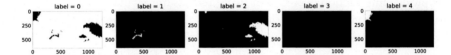

Figure 7.5 Class identification from 1D labels of the dataset.

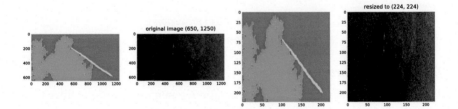

Figure 7.6 Resizing and pixel-wise class coloring of input image.

7.5.2.3 Data preprocessing

The provided dataset is preprocessed from SAR images to have contextual information for oil spill detection. Due to the functional altitude of the SAR sensor, the captured object might have a variable scale in representation, which might impose challenges in the training phase. Considering this issue, the authors of the dataset use a multi-scale scheme for image extraction from SAR images. One of the challenges in the oil spill detection process is to distinguish oil spills from the look-alike class, which has a similar representation in SAR images (black spots). This challenge gets more complicated with the ambiguous shape of the oil slick and in scenarios where a wide area of look-alike class belongs to the proximity of the oil slick.

7.5.3 Evaluation criteria

In semantic segmentation, the evaluation criteria of the DNN model are slightly different than the conventional accuracy measurement. Therefore, among other standard metrics for evaluating semantic segmentation models, Intersection over Union (IoU) and pixel accuracy were chosen for this work. The central concept of IoU and pixel accuracy is explained in the following subsections.

7.5.3.1 IoU

Typically, Intersection over Union (IoU) reflects the accurate identification of each class that belongs to the actual class. It is widely used in literature to evaluate semantic segmentation models. IoU reflects the intersection or overlap of pixels from predicted and true classes. From the mathematical point of view, IoU estimates the number of pixels common between the ground truth and prediction masks divided by the total number of pixels exiting across both masks that are defined in Eq. (7.2):

$$IoU = \frac{ground\ truth \cap prediction}{ground\ truth \cup prediction} \qquad (7.2)$$

7.5.3.2 Pixel accuracy

Another typical metric to evaluate semantic segmentation is estimating the percent of correctly classified pixels. It is commonly reported for each class individually and in overall classes. The main idea of pixel-wise accuracy is to evaluate a binary mask where true positive and true negative concepts are important factors. A true positive is the identification of a pixel that belongs to the true class. In contrast, true negative reflects the correct identification of a pixel that does not belong to the true class [203]. The pixel accuracy can be formulated in Eq. (7.3):

$$Pixel\ accuracy = \frac{TP + TN}{TP + TN + FP + FN} \qquad (7.3)$$

Pixel accuracy can provide deceiving information due to having fewer pixels for a class within an image. In this case, the estimation gets biased by addressing negative case identification (*i.e.*, the absence of a particular class in the image).

7.5.4 Experimental results

This research implemented the FCN-8s model to perform semantic segmentation on the oil spill dataset considering IoU and Pixel Accuracy as the main evaluation metric. To analyze the results, the authors focus on fine-tuning the implemented model's optimizers and learning rates among different hyperparameters. The case study considers the base model when using any optimizer with its default parameters. As the problem is multiclass semantic segmentation, the authors used categorical cross–entropy for compiling the FCN-8s model. At first, this work used a stochastic gradient descent (SGD) optimizer with its default values from Keras documentation and trained the model with 1002 images. The researchers split the train instances to have training and validation samples of 851 and 151, respectively. This work used 200 epochs to train the FCN-8s model with a batch size of 32. To test the FCN-8s model, 110 images are used that are provided in the oil spill dataset. From the trained model, researchers estimate the IoU for every class and estimate its mean. For the base model with default SGD, the mean IoU is around 46%. The detail result is provided in Table 7.3. Fig. 7.7 represents the training and validation loss for the base model, where the training loss is approximately 0.1, and the validation loss is around 0.2.

Table 7.3 Detail result of base model run with IoU and pixel accuracy using SGD optimizer default mode.

Class	IoU	Pixel accuracy
Sea Surface	94.50%	94.90%
Oil Spill	18.10%	98.80%
Look–alike	44.60%	96.30%
Ship	0.10%	100.0%
Land	76.00%	99.10%
mIoU	46.70%	97.82%

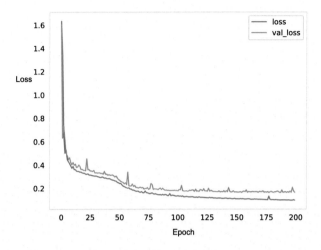

Figure 7.7 Training loss and validation loss for base model of FCN-8s.

7.5.5 Experiments with optimizers

To improve the performance of FCN-8s, the authors of this work experimented with utilizing different optimizers and changing the learning rates within each optimizer. As of the initial experimentation result, among five different optimizers, namely SGD, Adam, Adamax, Adadelta, and Agard, it is found that SGD, Adadelta, and Adamax perform better than Adam and Adagard. However, Adam and Agard optimizers perform worst while predicting classes from the test images. Accordingly, these two optimizers are discarded for further experimentation. In the following subsection, experiments with different optimizers are explained.

7.5.5.1 SGD optimizer

The implemented FCN-8s model has been trained with SGD optimizer by gradually changing the learning rate from 0.01 to 0.05 with a step size of 0.01 and estimating the mIoU for every trained model. The result was plotted in Fig. 7.8, where it demonstrates that for SGD with a learning rate of 0.05, the model has the best performance. Additionally, the increment of the learning rate improves the performance of the FCN-8s model in terms of mIoU, which is justified by the upward trend of the graph.

The detailed class IoU for the best model in terms of learning rate is demonstrated in Table 7.4. This table reflects the class level evaluation of the FCN-8s model, where the sea surface class has the highest IoU (94.60%), and the ship class has the lowest (2.30%). The result reflects the

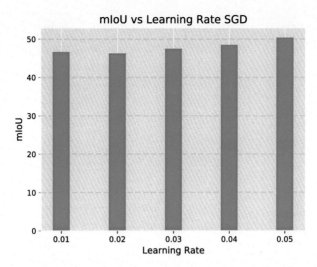

Figure 7.8 The changes of mIoU with respect to learning rate using SGD optimizer for training FCN-8s model.

Table 7.4 FCN-8s model evaluation utilizing SGD optimizer with learning rate 0.05.

Class	IoU	Pixel accuracy
Sea Surface	94.60%	95.00%
Oil Spill	29.30%	98.90%
Look–alike	43.20%	96.00%
Ship	2.30%	100.0%
Land	82.70%	99.40%
mIoU	50.40%	97.86%

class imbalance problem in the dataset, which is reported earlier in Table 7.1. Additionally, this table shows that Oil Spill class IoU improved to 29.30% from 18.10% reported from base model implementation. Therefore, using the SGD optimizer with a learning rate of 0.05, it was found that there was approximately 11% improvement in the Oil Spill detection class.

A loss graph was plotted against epochs to identify how the model converges (learning rate 0.05) for SGD. Fig. 7.9 demonstrates that for the loss function, the model converges rapidly within 30 epochs. However, after that model goes to saturation, and it can not reduce the loss any further. Finally, to visualize the FCN-8s model prediction on test samples, the true mask results were plotted and predicted in one image. Fig. 7.10 demon-

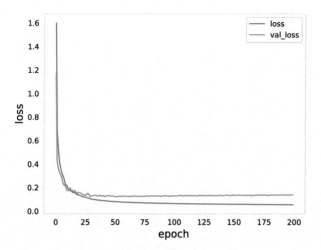

Figure 7.9 Training loss and validation loss for FCN-8s model while having learning rate 0.05.

strates the result of the prediction in terms of visual comparison for the highest mIoU FCN-8s model.

Fig. 7.10 reflects that the FCN-8s model with SGD optimizer performs reasonably for oil spill detection for the first row of images. The model confuses the oil spill with a look-alike in the second row. Finally, in the last row of Fig. 7.10, the model reasonably identifies look-alike from the test image.

7.5.5.2 Adadelta optimizer

In this experiment, the Adadelta optimizer is used for training the FCN-8s model, and fine-tuning has been performed with the learning rate. In this case, it is found that the Adadelta default learning rate is 1.0 for Keras' implementation. Therefore, the model is trained with the Adadelta optimizer using its default learning rate and estimates the IoU and Pixel accuracy, which is represented in Table 7.5.

For finding the optimal learning rate with the Adadelta optimizer, an experiment has been performed similarly to the first experiment. The experiment started with a 0.1 learning rate and gradually increased it up to 0.5. The mIoU is estimated for every change and plotted using a line plot to analyze the result. From Fig. 7.11, it is found that the highest performance in terms of mIoU is 50.45% for the Adadelta optimizer with a learning rate of 0.2. To further analyze the class level evaluation, the IoU for every class

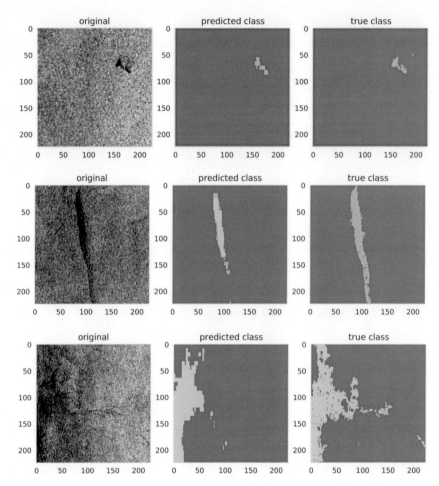

Figure 7.10 FCN-8s prediction on test image in terms of predicted class and true class for SGD optimizer.

is reported for this learning rate in Table 7.6. The class level comparison shows that, although overall performance improved in small amounts, the IoU for the Oil Spill class decreased from 31.40% to 29.00% while increasing the learning rate. Therefore, it is visible from the above experiments that with the Adadelta optimizer, the default learning rate performs better for the Oil Spill class. On the contrary, the overall performance (mIoU) is highest, with a learning rate of 0.2.

In the training of the FCN-8s model with Adadelta optimizer and optimal learning rate, the loss for every epoch is plotted in a graph that

Table 7.5 FCN-8s model evaluation utilizing Adadelta optimizer with default learning rate 1.0.

Class	IoU	Pixel accuracy
Sea Surface	94.80%	95.10%
Oil Spill	31.40%	99.00%
Look–alike	38.70%	96.20%
Ship	0.50%	100.00%
Land	84.80%	99.40%
mIoU	49.90%	97.94%

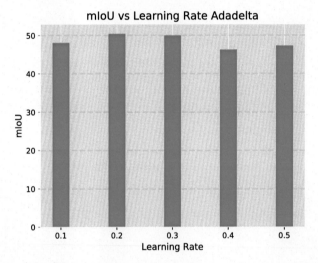

Figure 7.11 The changes of mIoU with respect to learning rate using Adadelta optimizer for training FCN-8s model.

represents the model convergence to loss. Fig. 7.12 represents the loss graph for Adadelta optimizer. It captures the convergence of the FCN–8s model in terms of loss update, where it is found that after 75 epochs, the model becomes saturated. Although the graph has some unusual spikes from 130 to 180 epochs, the overall loss updates last for a higher number of epochs than the SGD optimizer.

To visualize the prediction of the FCN–8s model, plot images have been made in Fig. 7.13 side by side for a better understanding of the model's performance. Fig. 7.13 reflects that in the first row of images, the oil spill is predicted with less IoU than SGD optimizer trained model. In the second row, the model again confuses with oil spill with a look–alike. Although, it

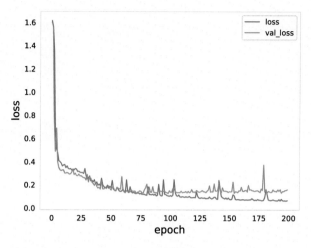

Figure 7.12 FCN-8s model training loss updates with respect to every epoch when Adadelta optimizer is used with 0.2 learning rate. This reflects the convergence of the model.

Table 7.6 FCN-8s model evaluation utilizing Adadelta optimizer with optimal learning rate 0.2.

Class	IoU	Pixel accuracy
Sea Surface	95.10%	95.40%
Oil Spill	29.00%	99.00%
Look-alike	41.10%	96.40%
Ship	1.00%	100.00%
Land	85.70%	99.50%
mIoU	50.45%	98.06%

performs better than the SGD optimizer for the second row of test image prediction. Finally, in the last row, the model identifies the look–alike class with less IoU than SGD.

7.5.5.3 Adamax optimizer

To understand how Adamax optimizer performs for training FCN-8s model on oil spill dataset, another experiment has been performed with learning rates similar to SGD and Adadelta optimizer. The Adamax has the default learning rate of 0.002, and training with this learning rate yields a significantly worse model of test accuracy, 16.60%, whereas loss goes up to

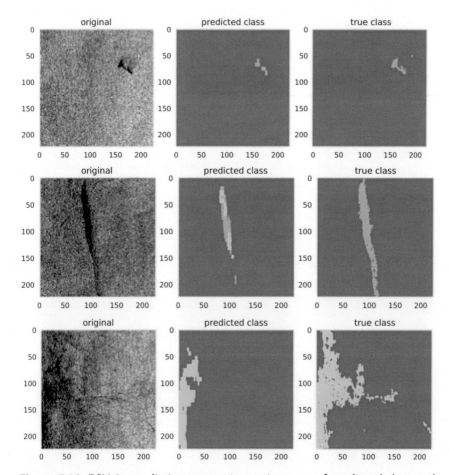

Figure 7.13 FCN-8s prediction on test image in terms of predicted class and true class for Adadelta optimizer.

13.44. Considering this incomparable result, the learning is reduced rate gradually, and the estimation of the mIoU for every training and testing has been performed. The result is plotted in a bar chart to understand the models' performance in terms of mIoU while changing the learning rate within the Adamax optimizer.

From Fig. 7.14, it is visible that for a learning rate of 0.001, the highest mIoU is observed. In this case, the optimal learning rate is 0.001, slightly less than the default learning rate. Therefore, reducing the learning rate can significantly improve the model's performance. The main concept inferred from this experiment is that while using Adamax optimizer, a reasonably

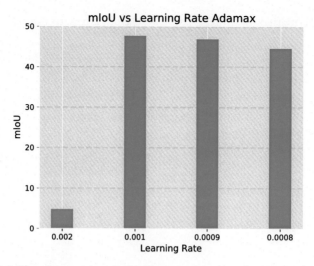

Figure 7.14 The changes in mIoU with respect to learning rate using Adamax optimizer.

Table 7.7 FCN-8s model evaluation utilizing Adamax optimizer with optimal learning rate 0.001.

Class	IoU	Pixel accuracy
Sea Surface	94.20%	94.50%
Oil Spill	30.30%	98.90%
Look-alike	31.70%	95.70%
Ship	0.10%	100.0%
Land	82.20%	99.30%
mIoU	47.70%	97.68%

low learning rate performs better than the default learning rate assigned by Keras API. Furthermore, the IoU for every class is estimated for a learning rate of 0.001 while using the Adamax optimizer to understand the class level accuracy.

Table 7.7 reflects the class level performance of the FCN-8s model while trained with Adamax optimizer and a learning rate of 0.001. From Table 7.7, it is understandable that the Oil Spill class has reasonable IoU, although overall performance (47.70%) is not good enough like the other two optimizers. Finally, to visualize the prediction of the oil spill and look-alike classes, a plot is made using the original image, prediction by FCN-8s model with Adamax optimizer, and true class in a row to estimate the performance.

Fig. 7.15, represents that for oil spill class (row 1 and 2), FCN-8s model trained with Adamax optimizer performs reasonably well as the true class (yellow) is identified properly. However, for the look–alike class (row 3), the model could identify fewer pixels than the true class. Here, the green represents the look–alike class.

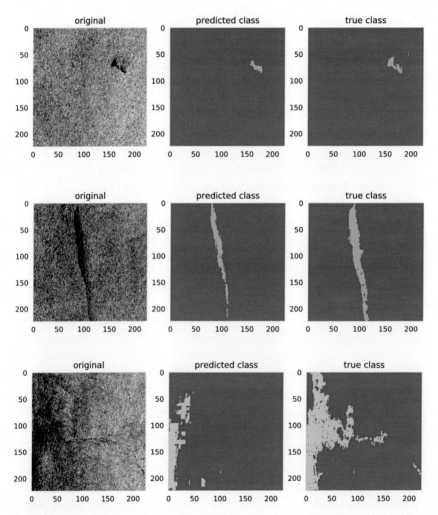

Figure 7.15 Prediction of Oil Spill and look-alike classes for Adamax optimizer with optimal learning rate of 0.001.

7.6 Summary

In this case study, the main goal is to implement a DNN model for multiclass semantic segmentation on the oil spill detection dataset and improve the model to some extent from its base implementation. Here, the base implementation of the model is defined as FCN-8s with an optimizer using default settings (*i.e.*, learning rate, decay, momentum). The authors utilize the FCN-8s model for this recent MKLab's published oil spill detection dataset, as the model has not been used on this dataset before. Hence, the FCN-8s model is implemented by fine-tuning different hyperparameters to improve both class level accuracy (mIoU) and overall performance. It is observed that the performance of the FCN-8s model can be improved by utilizing different optimizers with a particular configuration (*e.g.*, learning rate, momentum). Performing various experiments with optimizers and learning rates, the model could improve by approximately 4 to 5%, indicating that further fine-tuning can improve performance. Therefore, this work concludes the case study by suggesting a proper configuration of the DNN network model for implementing a robust oil spill detection system.

CHAPTER 8

Case study II: Evaluating DNN applications in smart O&G industry

8.1 Introduction

Software solutions operating based on machine learning and, particularly, Deep Neural Network (DNN) models are becoming fundamental pillars of Industry 4.0 revolution [204]. In the industrial automation process, numerous smart sensors frequently produce and fed data to the DNN-based applications that can make smart latency-sensitive decisions to improve energy efficiency, production, and safety measures. Building robust Industry 4.0 solutions entail having an accurate estimation of the inference (execution) time of DNN-based applications hosted on the cloud or edge computing systems. Lack of such assessments often leads to missing the applications' latency constraints and lowers their quality of service (QoS) [205] or increase the incurred cost of cloud resources. In critical industrial sectors, such as oil and gas, the penalty of such inaccurate estimations can be disastrous and cause unintended consequences, such as an unsafe workplace, environmental footprints, energy wastage, and damaging devices [4,206]. Accordingly, the goal in this study is to measure and model the stochasticity that exists in the execution time of industrial DNN-based applications that are commonly deployed in the cloud.

The primary motivation in this study is the critical industry of Oil and Gas (O&G) that is aimed at becoming clean and ultimately unmanned, thereby safe, in Industry 4.0. O&G is one of the main environmental pollutants and even minor improvements in this industry can have major impacts in the global scale. In this context, there are several time-sensitive operations (*e.g.*, fire detection [207] and toxic gas monitoring [208]) that failing to timely process them can potentially lead to disasters, such as oil spills, explosions, and even death.

Understanding the uncertainties exist in execution time of different application types and properly modeling them is crucial in architecting software solutions that are robust against these uncertainties. Note that DNN-based applications encompass both the training and inference stages

IoT for Smart Operations in the Oil and Gas Industry
https://doi.org/10.1016/B978-0-32-391151-1.00017-4

[209]. While the training stage is generally carried out offline, the focus in this study is on modeling the inference execution time that has to be accurately estimated for latency-sensitive and mission critical applications [210]. For instance, accurate estimation of inference time is instrumental in calculating the completion time of arriving tasks that can, in turn, help to make more precise resource allocation decisions [211].

Public or private Cloud datacenters, such as Amazon [212], are widely used as the back-end platform to execute DNN-based industrial applications [213]. The cloud providers often offer heterogeneous machine types, such as CPU-Optimized, Memory-Optimized, and GPU, that provide different execution time for various application types. For instance, a big data application type has its lowest execution time on the Memory-Optimized machine type whereas an image rendering application is best fitted to the GPU-based machine type. This form of heterogeneity is known as *inconsistent heterogeneity* [214,215]. For each machine type, cloud providers offer a *consistent heterogeneity* in form of various virtual machine (VM) instance types with different number of allocated resources. For example, in Amazon cloud, for CPU-Optimized machine type, there is *c5d.18xlarge* VM type with 36 number of cores that is faster than *c5.xlarge* VM type with only 2 cores. As each application type can potentially have different inference time on distinct machine types, it is critical to consider the resource heterogeneity in estimating the execution time of different DNN-based applications.

For that purpose, in this study, researchers evaluate and analyze the execution time of DNN-based applications on heterogeneous cloud machine types. The study encompasses both the application-centric perspective, by the way of modeling inference time, and the resource-centric perspective, by the way of measuring the Million Instruction Per Seconds (MIPS) metric. MIPS is considered as a rate-based metric that reflects the performance of cloud machine instance in terms of execution speed. As researchers consider latency-constrained applications, the underlying systems are considered as a dynamic (online) platform that processes each task upon arrival.

Prior studies on evaluating and modeling DNN-based applications [216] mostly focus on the core DNN model and ignore the end-to-end latency of the application that includes at least two other factors: (a) the latency of non-DNN parts of the application (*e.g.*, those for pre- and post-processing); and (b) the latency imposed due to uncertainties inherent to the cloud platform. Nonetheless, for critical industrial applications, such as those in O&G, a holistic analysis that considers the end-to-end latency of DNN-based applications is needed. The lack of such study hinders the path to develop a robust smart O&G solutions [180]. Accordingly, the main contributions of this work are as follows:

- Providing an application-centric analysis by developing a statistical model of the inference time of various DNN-based applications on heterogeneous cloud resources.
- Providing a resource-centric analysis of various DNN-based applications on heterogeneous cloud resources by developing a statistical model of MIPS, as a rate-based metric.
- Providing a publicly available[1] collection of pretrained DNN-based industrial applications in addition to their training and testing datasets. Moreover, a trace of inference execution times of the considered applications on heterogeneous machines of two public cloud platforms (namely AWS and Chameleon Cloud [217]) is presented.

8.2 DNN-based applications in O&G Industry 4.0

Table 8.1 summarizes different types of DNN-based applications used in the smart O&G industry. The table shows the abbreviated name for each application, its DNN (network) model, type of its input data, the scope of deployment in O&G Industry 4.0 [218], and the code base to build the model. All the applications, the input data, and analysis results are publicly available for reproducibility purposes in the Github repository mentioned earlier. In the rest of this section, we elaborate on the characteristics of each application type.

8.2.1 Fire detection (abbreviated as Fire)

Smart fire detection, a critical part of monitoring systems, aims at providing safety and robustness in Industry 4.0. Researchers analyzed a fire detection application developed by Dunnings and Breckon [207] using convolutional neural network (CNN). It automatically detects fire regions (pixels) in the frames of a surveilled video in a real-time manner. Among other implementations, researchers deploy the FireNet model that accurately identifies and locate fire in each frame of a given video segment. FireNet is a lightweight variation of AlexNet model [219] with three convolutional layers of sizes 64, 128, and 256. In this model, each convolutional layer is augmented

---

[1] https://github.com/hpcclab/Benchmarking-DNN-applications-industry4.0.

Table 8.1 DNN-based applications used in O&G Industry 4.0 along with their network model, input data type, usage scope, and code base.

Application type	DNN model	Input type	Scope	Code base
Fire Detection (Fire)	Customized AlexNet	Video Segment	Control & Monitoring	TensorFlow (tflearn)
Human Activity Recognition (HAR)	Customized Sequential Neural Network	Motion sensors	Workers Safety	Keras
Oil Spill Detec. (Oil)	FCN-8	SAR Images	Disaster Management	Keras
Acoustic Impedance Estimation (AIE)	Temporal Convolutional Network	Seismic Data	Seismic Exploration	PyTorch

by a max–pooling layer and a local response normalization to achieve high frequency features with a large response from previous layer.

8.2.2 Human activity recognition (abbreviated as HAR)

Human Activity Recognition (HAR) systems are widely used in Industry 4.0 to ensure workers safety in hazardous zones. For this purpose, motion sensors, such as accelerometer and gyroscope, that are widely available on handheld PDA devices are utilized. The HAR system operates based on the sequential neural network model with four layers to identify the worker's activities (namely, walking, walking upstairs, walking downstairs, sitting). For analysis, we use a dataset of UCI machine learning repository, known as Human Activity Recognition Using Smartphones [220].

8.2.3 Oil spill detection (abbreviated as Oil)

Detecting the oil spill is of paramount importance to have a safe and clean O&G Industry 4.0. The accuracy of DNN-based oil spill detection systems has been promising [186]. Researchers utilize a detection system that operates based on the FCN-8 model [198], which is depicted in Fig. 8.1. As we can see, the model contains five Fully Convolutional Network (FCN) blocks and two upsampling blocks that collectively perform semantic segmentation (*i.e.*, classifying every pixel) of an input image and output a labeled image. The FCN-8 model functions based on the satellite (a.k.a.

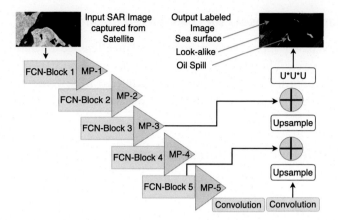

Figure 8.1 The FCN-8 model is presented in block diagram that consist of 5 fully convolutional network blocks, and 2 upsampling blocks. The model receives input as a SAR image and performs pixel-wise classification to output a labeled image.

SAR) [221] images. Researchers configure the analysis to obtain the inference time of 110 SAR images collected by MKLab [186].

8.2.4 Acoustic impedance estimation (abbreviated as AIE)

Estimating acoustic impedance (AI) from seismic data is an important step in O&G exploration. To estimate AI from seismic data, researchers utilize a solution functions based on the temporal convolutional network [222], shown in Fig. 8.2. The solution outperforms others in terms of gradient vanishing and overfitting. Marmousi 2 dataset [223] is used to estimate AI.

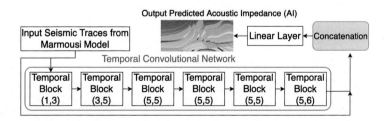

Figure 8.2 Schematic view of Temporal Convolutional Network (TCN) model that consists of six temporal blocks, the input data, and the output in form of the predicted AI.

8.3 Inference datasets for DNN-based applications

The DNN-based applications mainly perform inference operation based on input dataset. The inference operations of four different DNN-based applications mentioned in earlier section evaluated in this study with four different input datasets that are explained briefly in the following subsections.

8.3.1 Fire detection inference dataset

The fire detection application works based on FireNet DNN model that is trained with 23,408 images collected from two different literature works Chenebert et al. [224], and Steffens et al. [225] respectively. Additionally, public sources like "YouTube" are incorporated to make it randomized. For validation purpose, 2931 images are used with statistical evaluation. To explore the inference execution time evaluation this study utilizes a synthesized benchmark dataset of 240 videos with different frame rates, and resolutions. For fair and realistic analysis, the length of all videos is considered two seconds.

8.3.2 HAR inference dataset

Experiments on Human Activity Recognition were conducted with a group of 30 participants ranging in age from 19 to 48 years. Each participant wore a smartphone (Samsung Galaxy S II) around their waist and did six activities (WALKING, WALKING UPSTAIRS, WALKING DOWN-STAIRS, SITTING, STANDING, LAYING). Smartphones with inbuilt accelerometers and gyroscopes at a constant rate of 50 Hz collect 3-axial linear acceleration and 3-axial angular velocity. The tests were recorded so that the data could be manually labeled. The collected dataset was randomly partitioned into two sets, with 70% of the volunteers chosen to provide training data and 30% chosen to generate test data. In this study, this 30% of the data are used for inference operation.

8.3.3 Oil inference dataset

Satellite images captured with Synthetic Aperture Radar (SAR) technology are utilized to generate the oil spill detection dataset [186]. The dataset consists of 1112 images extracted from the raw SAR data, including the preliminary data to be manipulated. Five important classes are considered to be incorporated in the annotation: oil spills, look-alikes, ships, land, and sea surface. Each image is annotated using information from the EMSA records and human identification for maximum efficiency. Because the presented dataset was designed to be used with semantic segmentation methods, each of the five classes was assigned a different RGB color. As a result, the ground truth masks that accompany the dataset's images are colored according to the identified class for each instance of interest. The annotated dataset was divided into training (90%) and testing (10%), with 1002 and 110 images, respectively. This study uses 110 testing images for the inference operation.

8.3.4 AIE inference dataset

The Marmousi data [226] collection was created at the Institute de Francis du Petrol (IFP) in Paris, France, under the guidance of Roloff Versteeg (1990). The velocity of Earth model is based on a real prospect discovered off the coast of Western Africa. The Marmousi data [227] have proven to be a rich mine for both developing image algorithms and demonstrating to practitioners of the art that the world is not as simple as they once believed. The Marmousi2 dataset [228] is a more flexible version of the original Marmousi model that was designed by Allied Geophysical Laboratories (AGL). The Marmousi2 model has been widely used in amplitude versus offset (AVO) analysis, impedance inversion, multiple attenuation, and multicomponent imaging, among other applications. In addition, AGL has made the data available to researchers worldwide.

Throughout this model, three sets of data were gathered. A near-surface streamer survey, a vertical sounding profile (VSP), and an ocean bottom cable (OBC) survey are all part of the project. In addition, the Marmousi2 repository contains many sets of short records that are used for DNN model training. This study utilizes a portion of these data that are not used in training for the inference operation.

8.4 Cloud computing platforms for Industry 4.0

8.4.1 Amazon cloud

AWS is a pioneer in the Cloud computing industry and provides more than 175 services, including Amazon EC2 [229], across a large set of distributed data centers. Amazon EC2 provides inconsistently heterogeneous machines (*e.g.*, CPU, GPU, and Inferentia) in form of various VM instance types (*e.g.*, general purpose, compute-optimized, and machine learning (ML)). Within each VM type, a range of VM configurations (*e.g.*, large, xlarge, 2xlarge) are offered that reflect the consistent heterogeneity within that

VM type. To realize the impact of machine heterogeneity on the inference time of various applications, researchers choose one representative VM type of each offered machine type. Table 8.2 represents the type of machines and their specification in terms of number of cores and memory. This is to note that all the machine types use SSD storage units. Although General Purpose machines are not considered suitable for latency-sensitive DNN-based applications, the reason researchers study them is their similarity to the specifications of machine types often used in the edge computing platforms. As such, considering these types of machines (and similarly *m1.small* in the Chameleon cloud) makes the results of this study applicable to cases where edge computing is employed for latency-sensitive applications [108].

Table 8.2 Heterogeneous machine types and VM configurations in Amazon EC2 that are considered for performance modeling of DNN-based applications. In this table, ML Optimized represents Inferentia machine type offered by AWS.

Machine type	VM config.	vCPU	GPU	Mem. (GB)
Mem. Optimized	r5d.xlarge	4	0	32
ML Optimized	infl.xlarge	4	0	8
GPU	g4dn.xlarge	4	1	16
General Purpose	m5ad.xlarge	4	0	16
Comp. Optimized	c5d.xlarge	4	0	8

Table 8.3 Various VM flavors in Chameleon cloud are configured to represent a consistently heterogeneous system.

VM config.	vCPUS	Mem. (GB)
m1.xlarge	8	16
m1.large	4	8
m1.medium	2	4
m1.small	1	2

8.4.2 Chameleon cloud

Chameleon cloud [217] is a large-scale public cloud maintained by National Science Foundation (NSF). Chameleon cloud supports VM-based heterogeneous computing. It offers the flexibility to manage the compute, memory, and storage capacity of the VM instances. In this study, researchers use the Chameleon cloud to configure a set of consistently heterogeneous machines. Researchers configure four VM flavors, namely `m1.xlarge, m1.large, m1.medium,` and `m1.small`, as detailed in Table 8.3. This is to note that VMs offered by Chameleon cloud use HDD unit as storage.

8.5 Performance modeling of inference time

The focus of this study is on latency–sensitive DNN–based applications that are widely used in Industry 4.0. Accordingly, this work consider a dynamic (online) system that is already loaded with pretrained DNN–based applications, explained in the previous section, and executes arriving requests on the pertinent application. This means that this work measures the hot start inference time [230] in the considered applications. The DNN–based applications mostly use TensorFlow, and the VMs both in AWS and Chameleon are configured to use the framework on top of Ubuntu 18.04.

Figure 8.3 The stochastic nature of inference execution time of oil spill application while running on heterogeneous VMs in the AWS. For every VM instance, the oil spill detection application is executed 30 times and those executions are plotted as number of attempts along x-axis. The y-axis represents the inference time for every attempt.

The initial evaluations in AWS (shown in Fig. 8.3) demonstrate that, in different attempts, the inference execution time of an application (Oil Spill) on the same machine type can be highly stochastic. Similar stochasticity is found for chameleon cloud while running the oil spill detection application 30 times within same VM instance. Hence to capture this randomness (aka consistent heterogeneity) that is caused by several factors, such as transient failures or multi–tenancy [231,232], the analysis is based on 30 times execution of the same request within same VM.

8.5.1 Application-centric analysis of inference time
8.5.1.1 Overview

In this part, researchers capture the inference time of the four DNN applications and try to identify their underlying statistical distributions using various statistical methods. Then, to describe the behavior of inference execution time using a single metric, they explore the central tendency of the distributions.

8.5.1.2 Statistical distribution of inference execution time

Among various statistical methods, the normality tests are widely employed to understand the distribution of the collected samples. Hence, researchers first use Shapiro-Wilk test [233] to verify the normality of the inference time distribution on each machine type. Next, they employ Kolmogorov-Smirnov test [234] to find the best fit distribution based on the sampled inference execution times.

8.5.1.3 Shapiro-Wilk test to verify normality of the sampled data

The null hypothesis is that the inference execution times are normally distributed. To understand whether a random sample comes from a normal distribution, researchers perform the Shapiro-Wilk test. The result of this test is considered as W, whose low value (lower than w_α threshold) indicates that the sampled data are not normally distributed and vice versa. The value of W is used to perform the significant testing (*i.e.*, calculating P-value). The higher P-value, especially greater than a threshold value (typically 0.05), supports the null hypothesis that the sampled data are normally distributed.

The results of Shapiro-Wilk test on the collected inference times for AWS are presented in Table 8.4, where columns present the various machine types and rows define the application types. The table reflects that the initial assumption is not totally valid. The Shapiro-Wilk test result for the Chameleon cloud, depicted in Table 8.5, shows that for only three of the total cases, the normality assumption holds. Considering the lack of normality in several cases, in the next section, researchers utilize Kolmogorov-Smirnov test to more granularly explore the best fitting distribution for the inference time of each application and also cross validate the prior results obtained using another statistical method.

8.5.1.4 Kolmogorov-Smirnov goodness of fit test

The null hypothesis for the Kolmogorov-Smirnov test is that the inference times of a certain application type on a given machine type follow a statistical distribution. The Kolmogorov-Smirnov Goodness of Fit test (a.k.a. *K-S test*) identifies whether a set of samples derived from a population fits to a specific distribution. Precisely, the test measures the largest vertical distance

Table 8.4 The execution time distributions of DNN-based applications in AWS clouds machines using Shapiro-Wilk test.

App. Type	Mem. Opt.	ML Opt.	GPU	Gen. Pur.	Compt. Opt.
		Execution time distribution with Shapiro-Wilk test in AWS cloud			
Fire	Not Gaussian ($P = 2.73e^{-16}$)	Not Gaussian ($P = 5.42e^{-16}$)	Not Gaussian ($P = 6.59e^{-16}$)	Not Gaussian ($P = 2.06e^{-13}$)	Not Gaussian ($P = 3.9e^{-16}$)
HAR	Not Gaussian ($P = 7.12e^{-8}$)	Not Gaussian ($P = 1.04e^{-8}$)	Gaussian ($P = 0.19$)	Not Gaussian ($P = 1.76e^{-8}$)	Not Gaussian ($P = 0.4.62e^{-5}$)
Oil	Not Gaussian ($P = 8e^{-4}$)	Not Gaussian ($P = 2.9e^{-16}$)	Not Gaussian ($P = 0.012$)	Not Gaussian ($P = 1.27e^{-16}$)	Not Gaussian ($P = 5.86e^{-14}$)
AIE	Gaussian ($P = 0.46$)	Gaussian ($P = 0.23$)	Gaussian ($P = 0.08$)	Not Gaussian ($P = 1.99e^{-10}$)	Gaussian ($P = 0.96$)

Table 8.5 The execution time distributions of DNN applications in Chameleon cloud using Shapiro-Wilk test.

App. Type	Execution time distribution with Shapiro-Wilk test in Chameleon			
	m1.xlarge	m1.large	m1.medium	m1.small
Fire	Not Gaussian $(P = 4.05e^{-5})$	Not Gaussian $(P = 1.e^{-4})$	Not Gaussian $(P = 7.58e^{-6})$	Not Gaussian $(P = 1.32e^{-7})$
HAR	Gaussian $(P = 0.74)$	Not Gaussian $(P = 0.02)$	Gaussian $(P = 0.18)$	Gaussian $(P = 0.84)$
Oil	Not Gaussian $(P = 0.01)$	Not Gaussian $(P = 5.5e^{-7})$	Not Gaussian $(P = 0.01)$	N/A
AIE	Not Gaussian $(P = 2.77e^{-10})$	Not Gaussian $(P = 3.46e^{-6})$	Not Gaussian $(P = 1.4e^{-4})$	Not Gaussian $(P = 2.46e^{-6})$

(called test statistic D) between a known hypothetical probability distribution and the distribution generated by the observed inference times (a.k.a. empirical distribution function (EDF)). Then, if D is greater than the critical value obtained from the K-S test P-Value table, then the null hypothesis is rejected.

The results of the K-S test on the observed inference times in AWS and Chameleon clouds are depicted in Tables 8.6 and 8.7, respectively. From Table 8.6, we find that, in AWS, majority of the entries either represent Normal distribution or Student's t-distribution that exposes similar properties. However, we observe that the inference time of Fire Detection application does not follow any particular distribution with an acceptable P-Value. We also observe that the inference times of both Oil Spill application on Compute Optimized machine and AIE application on General Purpose machine follow exponential distribution. However, low P-Value in both of these cases indicate a weak acceptance of the null hypothesis.

On the contrary, Table 8.7 reflects the dominance of Log-normal (*i.e.*, the logarithm of the random variable is normally distributed) and Student's t-distribution over other distributions in the Chameleon cloud. Analyzing the execution traces shows that the inference times in Chameleon are predominantly larger than the ones in AWS that causes right-skewed property, hence, the distribution tends to Log-normal. Consistent to AWS observations, we see that Fire Detection application, in most of the cases, does not follow any distribution. Further analysis showed that the lack of distribution is because of variety (*e.g.*, frame rate and resolution) in the input videos. In fact, when researchers reduced the degree of freedom in the in-

Table 8.6 Inference time distributions of DNN-based applications in AWS cloud machines using Kolmogorov-Smirnov test.

| App. Type | Mem. Opt. | ML Opt. | Execution time distribution with Kolmogorov-Smirnov test in AWS cloud | | | |
			GPU	Gen. Pur.	Compt. Opt.
Fire	No Distr.	No Distr.	No Distr.	No Distr.	No Distr.
HAR	Student's t (P = 0.08)	Student's t (P = 0.77)	Student's t (P = 0.99)	Student's t (P = 0.57)	Student's t (P = 0.95)
Oil	Student's t (P = 0.44)	Student's t (P = 0.96)	Student's t (P = 0.5)	Student's t (P = 0.20)	Exponential (P = 0.21)
AIE	Normal (P = 0.99)	Normal (P = 0.54)	Normal (P = 0.47)	Exponential (P = 0.16)	Normal (P = 0.99)

Table 8.7 Inference time distributions of DNN-based applications in Chameleon's machines using the K-S test.

Execution time distribution with Kolmogorov-Smirnov test in Chameleon				
App. Type	m1.xlarge	m1.large	m1.medium	m1.small
Fire	No Distr	No Distr	No Distr	Log-normal
HAR	Normal (P = 0.98)	Student's t (P = 0.88)	Normal (P = 0.66)	Normal (P = 0.96)
Oil	Log-normal (P = 0.36)	Log-normal (P = 0.99)	Log-normal (P = 0.81)	N/A
AIE	Student's t (P = 0.47)	Student's t (P = 0.12)	Student's t (P = 0.25)	Student's t (P = 0.83)

put videos limited them to those with the same configuration (frame-rate), they noticed the inference time follows a Log-normal distribution. The observation shows that the characteristics and variation of input data can be decisive on the statistical behavior of inference times (mentioned in highlighted insight). This is to note that Oil Spill application cannot be run on `m1.small` machine owing to its limited memory.

Insights: The key insights of the analysis conducted on identifying the distribution of inference time are as follows:

- Shapiro-Wilk test for AWS and Chameleon rejects the null hypothesis that the inference times of DNN-based applications follow the Normal distribution.
- The K-S test reflects the dominance of Student's t-distribution of inference time in both AWS (Table 8.6), and Chameleon (Table 8.7).
- Various configurations of input data are decisive on the statistical distribution of the inference time.

8.5.1.5 Analysis of central tendency and dispersion measures

Leveraging the statistical distributions of inference times, in this part, researchers explore their central tendency metric that summarizes the behavior of collected observations in a single value. In addition, to analyze the statistical dispersion of the observations, researchers measure the standard deviation of the inference times. In particular, they estimate the arithmetic mean and standard deviation of the inference times. The central tendency metrics of inference times for Chameleon and AWS clouds are shown in Tables 8.8 and 8.9, respectively. The **key insights** are as follows:

Table 8.8 Central tendency metric (μ), and data dispersion metric (σ) of the inference times in the Chameleon cloud.

Mean and standard deviation of inference execution times (ms) in Chameleon				
App. Type	m1.xlarge	m1.large	m1.medium	m1.small
Fire	$\mu = 2155.20$ $\sigma = 725.48$	$\mu = 2213.30$ $\sigma = 731.50$	$\mu = 2330.80$ $\sigma = 742.20$	$\mu = 3184.80$ $\sigma = 1033.30$
HAR	$\mu = 22.14$ $\sigma = 0.76$	$\mu = 47.69$ $\sigma = 1.26$	$\mu = 49.24$ $\sigma = 0.57$	$\mu = 52.69$ $\sigma = 0.78$
Oil	$\mu = 147.16$ $\sigma = 5.23$	$\mu = 222.22$ $\sigma = 2.89$	$\mu = 412.78$ $\sigma = 4.99$	N/A
AIE	$\mu = 6.23$ $\sigma = 0.25$	$\mu = 6.23$ $\sigma = 0.15$	$\mu = 6.40$ $\sigma = 0.13$	$\mu = 7.72$ $\sigma = 0.24$

- Machine Learning Optimized and GPU instances often outperform other AWS machine types.
- In both clouds, the inference times of Fire and Oil experience a higher standard deviation in compare with other applications. The high uncertainty is attributed to the characteristics of their input data; Oil Spill input images suffer from class imbalance [186], whereas, Fire input videos are highly variant.
- In Chameleon VMs with a consistent heterogeneity, DNN applications with dense network models (*e.g.*, Oil and Fire) can take advantage of powerful machines (*e.g.*, m1.xlarge) to significantly reduce their inference times.
- Overall, AWS offers a lower inference time than Chameleon. The reason is utilizing SSD units in AWS rather than HDD in Chameleon. In addition, it is noticed that Chameleon experiences more transient failures that can slow down the applications.

8.5.2 Resource-centric analysis of inference time

In performance analysis of computing systems, a rate-based metric [235] is defined as the normalization of number of computer instructions executed to a standard time unit. MIPS is a popular rate-based metric that allows comparison of computing speed across two or more computing systems. Given that computing systems (*e.g.*, AWS ML Optimized and GPU) increasingly use instruction-level facilities for ML applications, the objective in this part is to analyze the performance of different machine types in

Table 8.9 The measurement of central tendency metric (μ), and data dispersion metric (σ) on the observed inference times in AWS.

App. Type	Mem. Opt.	ML Opt.	GPU	Gen. Pur.	Compt. Opt.
	Mean and standard deviation of inference execution times (ms) in AWS				
Fire	$\mu = 1461.8$ $\sigma = 457.3$	$\mu = 1281.7$ $\sigma = 387.93$	$\mu = 1349.5$ $\sigma = 418.9$	$\mu = 1534.8$ $\sigma = 494.7$	$\mu = 1421.4$ $\sigma = 441.8$
HAR	$\mu = 1.27$ $\sigma = 0.082$	$\mu = 0.66$ $\sigma = 0.006$	$\mu = 0.51$ $\sigma = 0.006$	$\mu = 1.17$ $\sigma = 0.042$	$\mu = 0.66$ $\sigma = 0.003$
Oil	$\mu = 269.9$ $\sigma = 1.01$	$\mu = 218.8$ $\sigma = 0.66$	$\mu = 65.98$ $\sigma = 0.47$	$\mu = 667.1$ $\sigma = 2.26$	$\mu = 242.9$ $\sigma = 0.68$
AIE	$\mu = 7.02$ $\sigma = 0.02$	$\mu = 6.41$ $\sigma = 0.03$	$\mu = 7.55$ $\sigma = 0.04$	$\mu = 9.35$ $\sigma = 0.06$	$\mu = 7.95$ $\sigma = 0.02$

processing DNN-based applications. The results of this analysis can be of particular interest to researchers and cloud solution architects whose endeavor is to develop tailored resource allocation solutions for Industry 4.0 use cases. As for rate-based metrics we do not assume any distribution [236], we conduct a nonparametric approach. In addition to MIPS, we provide the range of MIPS in form of *Confidence Intervals* (CI) for each case.

Table 8.10 MIPS values of heterogeneous machines in AWS for each DNN-based application.

The MIPS for DNN applications in AWS cloud					
App. Type	Mem. Opt.	ML Opt.	GPU	Gen. Pur.	Compt. Opt.
Fire	1938.63	2196.35	2092.72	1862.04	1989.56
HAR	838640.65	1595874.34	2040057.33	891754.48	1581709.12
Oil	164.54	168.58	331.98	20.46	162.01
AIE	145.58	180.28	150.25	131.25	160.32

Table 8.11 MIPS vales for heterogeneous machines on Chameleon cloud for each DNN-based application.

The MIPS for DNN applications in Chameleon				
App. Types	m1.xlarge	m1.large	m1.medium	m1.small
Fire	1327.81	1282.33	1249.63	871.36
HAR	91.78	102.51	124.76	136.62
Oil	18267.35	11233.41	6243.94	N/A
AIE	246366.52	249551.29	236300.93	201807.49

Let application i with n_i instructions have t_{im} inference time on machine m. Then, MIPS of machine m to execute the application is defined as $MIPS_{mi} = n_i/(t_{im} \times 10^6)$. Hence, before calculating MIPS for any machine, it is necessary to estimate the number of instructions (n) of each DNN-based application. For that purpose, researchers execute each task t_i on a machine whose MIPS is known and estimated n_i. Then, for each machine m, they measure t_{im} and subsequently calculate $MIPS_{mi}$. Tables 8.10 and 8.11 show the MIPS values for AWS and Chameleon, respectively.

To measure the confidence intervals (CI) of MIPS for each application type in each machine type, researchers use the nonparametric statistical methods [236] that perform prediction based on the sample data without making any assumption about their underlying distributions. Considering the rate-based metric, researchers use harmonic mean that offers a precise analysis for this type of metric rather than the arithmetic mean. In this case, they utilize Jackknife [236] resampling method and validate it using Bootstrap [236], which is another well-known resampling method. Both of these methods employ harmonic means to measure the confidence intervals of MIPS.

8.5.2.1 Estimating confidence interval using Jackknife method

Let p be the number of observed inference times. The Jackknife method calculates the harmonic mean in p iterations, each time by eliminating one sample. That is, each time it creates a new sample (resample) with size $p - 1$. Let x_j be the jth observed inference time. Then, the harmonic mean of resample i is called the pseudo–harmonic value (denoted as y_i) and is calculated based on Eq. (8.1):

$$y_i = \frac{p - 1}{\sum_{j=1, j \neq i}^{p} \frac{1}{x_j}} \tag{8.1}$$

Next, the arithmetic mean (denoted $\bar{y}$) of the p pseudo–harmonic values is computed, and is used to estimate the standard deviation. Finally, the t-distribution table is used to calculate the CI boundaries with a 95% confidence level. The result of the Jackknife method for AWS machines is shown in Table 8.12 that conforms with the MIPS calculation in Table 8.10. Similarly, the results of analysis for Chameleon cloud using Jackknife method, shown in Table 8.13, validate the prior MIPS calculations in Table 8.11. However, in the next part, researchers cross-validate these results using Bootstrap method.

8.5.2.2 Estimating confidence interval using Bootstrap method

Bootstrap repeatedly performs random sampling with a replacement technique [236] on the observed inference times. The random sampling refers to the selection of a sample with the chance of nonzero probability and the number (represented as k) of resample data depends on the user's consideration. After resampling, the harmonic means of k number of samples are calculated and sorted in ascending order to estimate the confidence intervals. Finally, for a specific confidence level, the $(\alpha/2 \times k)$th and $((1 - \alpha/2) \times k)$th values are selected from the sorted samples as the lower and upper bounds of the CI. the k value is set to 100 and α to 0.05 for 95% confidence level.

For both AWS and Chameleon, the results of CI analysis using the Bootstrap method are similar to, thus validate, the ranges estimated by the

Table 8.12 The confidence intervals of MIPS values for DNN-based applications in AWS machines, resulted from Jackknife resampling method.

CI of MIPS using Jackknife method in AWS cloud

App. Type	Mem. Opt.	ML Opt.	GPU	Gen. Pur.	Compt. Opt.
Fire	[1549.42, 1975.65]	[1770.81, 2243.04]	[1671.78, 2131.66]	[1465.31, 1889.77]	[1594.78, 2028.36]
HAR	[812040.26, 856355.96]	[1592214.75, 1599426.64]	[2033084.47, 2046727.57]	[880417.69, 901345.49]	[1580275.10 1585598.85]
Oil	[163.55, 165.47]	[168.36, 168.81]	[330.68, 333.22]	[20.35, 20.57]	[161.86, 162.17]
AIE	[139.02, 141.04]	[155.56, 156.01]	[141.57, 142.03]	[118.06, 119.82]	[148.35, 149.00]

Table 8.13 Confidence intervals of MIPS values for different DNN-based applications in Chameleon machines, resulted from Jackknife resampling method.

App. Type	CI of MIPS using Jackknife method in Chameleon cloud			
	m1.xlarge	m1.large	m1.medium	m1.small
Fire	[1032.11, 1341.75]	[1010.62, 1303.02]	[964.76, 1259.68]	[670.82, 872.85]
HAR	[88.27, 94.20]	[99.84, 104.49]	[122.33, 126.67]	[135.13, 137.92]
Oil	[18083.59, 18628.64]	[11159.71, 11662.41]	[6139.59, 6262.15]	N/A
AIE	[237710.12, 252686.82]	[247166.73, 251673.68]	[168804.58, 268273.11]	[199676.71, 203681.17]

Jackknife method. Therefore, due to the shortage of space, researchers do not report the table of MIPS values for the Bootstrap method. However, this is to note that the CI ranges provided by the Bootstrap method are shorter (*i.e.*, have less uncertainty), regardless of the application type and the cloud platform. The reason for the shorter range is that Bootstrap performs resampling with a user-defined number of samples that can be larger than the original sample size.

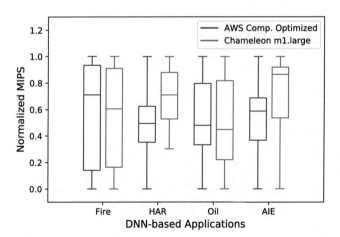

Figure 8.4 Comparative analysis of the MIPS values of AWS and Chameleon machines for various DNN-based applications. For the sake of presentation, the MIPS values are normalized between [0, 1].

To perform a cross-platform analysis of the MIPS values, in Fig. 8.4, researchers compare the range of MIPS values for AWS `Compute Optimized` against `m1.large` that is a compatible machine type in Chameleon (see Tables 8.2 and 8.3). The horizontal axis of this figure shows different application types, and the vertical axis shows the MIPS values, normalized based on MinMax Scaling in the range of [0, 1], for the sake of better presentation. Due to high variation in the input videos, it is observed that a broad CI range for Fire detection across both cloud platforms. However, for HAR, Oil Spill, and AIE applications, the observation is that the first and third quartiles of the CI range in Chameleon (whose machines are prone to more transient failures [237]) are larger than those in AWS. This wide range indicates that, apart from variations in the input data, the reliability of underlying resources is also decisive on the stochasticity of the inference times.

8.6 Summary and discussion

Accurately estimating the inference time of latency-sensitive DNN-based applications plays a critical role in robustness and safety of Industry 4.0. Such accurate estimations enable cloud providers and solution architects to devise resource allocation and load balancing solutions that are robust against uncertainty exists in the execution time of DNN-based applications. This work provides application- and resource-centric analyses on the uncertainty exists in the inference times of several DNN-based applications deployed on heterogeneous machines of two computing platforms, namely AWS and Chameleon. In the first part, the Shapiro-Wilk test is utilized to verify if the assumption of Normal distribution for the inference time holds. We observed that the inference times often do not follow a Normal distribution. Therefore, in the second part, researchers broaden the distribution testing investigation and utilized the Kolmogorov-Smirnov test to verify the underlying distributions in each case. The analysis showed that inference times across the two cloud platforms often follow Student's t-distribution. However, in several cases in Chameleon cloud we observed the Log-normal distribution that we attribute it to the uncertain performance of VMs in this platform. Next, to conduct a resource-centric analysis, researchers modeled MIPS (as a rate-based performance metric) of the heterogeneous machines for each application type. In the analysis, they took a nonparametric approach, which is suitable for rate-based metrics, and utilized the Jackknife and Bootstrap resampling methods with harmonic mean to determine the range of confidence intervals of the MIPS values in each case. The calculated MIPS values and their CI ranges reflect the behavior of different DNN-based applications under various machine types and cloud platforms. A comparative analysis of the CI ranges across AWS and Chameleon cloud demonstrates that the uncertainty in the inference time is because of variations in the input data and unreliability of the underlying platforms. In the future, researchers plan to incorporate the findings of this research to devise accurate resource allocation methods in IoT and edge computing systems. In addition, researchers plan to develop a predictive analysis to determine the execution of each inference task upon arrival.

Bibliography

[1] T.R. Wanasinghe, R.G. Gosine, L.A. James, G.K. Mann, O. de Silva, P.J. Warrian, The internet of things in the oil and gas industry: A systematic review, IEEE Internet of Things Journal 7 (9) (2020) 8654–8673.

[2] P. Flichy, C. Baudoin, et al., The industrial IoT in oil & gas: Use cases, in: SPE Annual Technical Conference and Exhibition, 2018.

[3] C. Toma, M. Popa, IoT security approaches in oil & gas solution industry 4.0, Informatica Economica 22 (3) (2018) 46–61.

[4] R. Hussain, M. Amini, A. Kovalenko, Y. Feng, O. Semiari, Federated edge computing for disaster management in remote smart oil fields, in: 2019 IEEE 21st International Conference on High Performance Computing and Communications; IEEE 17th International Conference on Smart City; IEEE 5th International Conference on Data Science and Systems (HPCC/SmartCity/DSS), IEEE, 2019, pp. 929–936.

[5] Wired | what is cloud?, https://www.accenture.com/_acnmedia/PDF-145/Accenture-What-is-Cloud-WIRED-Brand-Lab-Video-Transcript.pdf#zoom=50. (Accessed 29 April 2021).

[6] Cloud computing in the oil and gas industry | COPAS, Aug. 10, 2021. (Online; accessed 1 May 2022).

[7] Repsol launches big data, AI project at Tarragona refinery. (Online; accessed 30 April 2022).

[8] M. Polese, R. Jana, V. Kounev, K. Zhang, S. Deb, M. Zorzi, Machine learning at the edge: A data-driven architecture with applications to 5G cellular networks, IEEE Transactions on Mobile Computing (2020).

[9] J. Kim, H. Zeng, D. Ghadiyaram, S. Lee, L. Zhang, A.C. Bovik, Deep convolutional neural models for picture-quality prediction: Challenges and solutions to data-driven image quality assessment, IEEE Signal Processing Magazine 34 (6) (2017) 130–141.

[10] B. Chen, L. Liu, M. Sun, H. Ma, IoTCache: Toward data-driven network caching for internet of things, IEEE Internet of Things Journal 6 (6) (2019) 10064–10076.

[11] J. Wang, C. Jiang, K. Zhang, X. Hou, Y. Ren, Y. Qian, Distributed q-learning aided heterogeneous network association for energy-efficient IIoT, IEEE Transactions on Industrial Informatics 16 (4) (2020) 2756–2764.

[12] L.D. Xu, L. Duan, Big data for cyber physical systems in industry 4.0: a survey, Enterprise Information Systems 13 (2) (2019) 148–169.

[13] Y. Lu, Industry 4.0: A survey on technologies, applications and open research issues, Journal of Industrial Information Integration 6 (2017) 1–10.

[14] L. Zhou, D. Wu, J. Chen, Z. Dong, When computation hugs intelligence: Content-aware data processing for industrial IoT, IEEE Internet of Things Journal 5 (3) (2018) 1657–1666.

[15] J. Nilsson, F. Sandin, Semantic interoperability in industry 4.0: Survey of recent developments and outlook, in: 2018 IEEE 16th International Conference on Industrial Informatics (INDIN), 2018, pp. 127–132.

[16] K. Rao, G. Coviello, W. Hsiung, S. Chakradhar, ECO: Edge-cloud optimization of 5G applications, in: Proceedings of the 21st IEEE International Symposium on Cluster, Cloud and Internet Computing (CCGrid), May 2021, pp. 649–659.

[17] M.C. Lucas-Estañ, J. Gozalvez, Load balancing for reliable self-organizing industrial IoT networks, IEEE Transactions on Industrial Informatics 15 (9) (2019) 5052–5063.

[18] H. Pei Breivold, Towards factories of the future: migration of industrial legacy automation systems in the cloud computing and internet-of-things context, Enterprise Information Systems 14 (4) (2020) 542–562.

[19] M.S. Mahmoud, Architecture for cloud-based industrial automation, in: Third International Congress on Information and Communication Technology, 2019, pp. 51–62.

[20] T. Lewandowski, D. Henze, M. Sauer, J. Nickles, B. Bruegge, A software architecture to enable self-organizing, collaborative IoT resource networks, in: 2020 Fifth International Conference on Fog and Mobile Edge Computing (FMEC), 2020, pp. 70–77.

[21] Why digital twin technology is the future of IoT? – SnapStack, May 28, 2021. (Online; accessed 26 April 2022).

[22] A. Parrott, L. Warshaw, Industry 4.0 and the digital twin, Deloitte University Press, 2017, pp. 1–17.

[23] F. Tao, J. Cheng, Q. Qi, M. Zhang, H. Zhang, F. Sui, Digital twin-driven product design, manufacturing and service with big data, The International Journal of Advanced Manufacturing Technology 94 (9) (2018) 3563–3576.

[24] The increasing popularity of digital twins in oil and gas | GEP, Mar. 18, 2016. (Online; accessed 26 April 2022).

[25] Transform oil and gas with an intelligent edge and IoT, https://www.ciosummits.com/SB_Oil__Gas_IoT-1.pdf. (Accessed 25 April 2021).

[26] J.-H. Bae, D. Yeo, D.-B. Yoon, S.W. Oh, G.J. Kim, N.-S. Kim, C.-S. Pyo, Deep-learning-based pipe leak detection using image-based leak features, in: 2018 25th IEEE International Conference on Image Processing (ICIP), IEEE, 2018, pp. 2361–2365.

[27] T.R. Wanasinghe, L. Wroblewski, B.K. Petersen, R.G. Gosine, L.A. James, O. De Silva, G.K. Mann, P.J. Warrian, Digital twin for the oil and gas industry: Overview, research trends, opportunities, and challenges, IEEE Access 8 (2020) 104175–104197.

[28] K. Chowdhury, A. Arif, M.N. Nur, O. Sharif, A cloud-based computational framework to perform oil-field development & operation using a single digital twin platform, in: Offshore Technology Conference, OnePetro, 2020.

[29] How BP and Chevron use digital twins to optimize assets, https://bit.ly/3KgYkgn, 2022.

[30] Managing our assets, https://www.chevron.com/technology/managing-our-assets, 2020.

[31] S.K. Haldar, Mineral Exploration: Principles and Applications, Elsevier, 2018.

[32] H. Zhuang, Y. Han, X. Liu, H. Sun, Dynamic Well Testing in Petroleum Exploration and Development, Elsevier, 2020.

[33] H. Crumpton, Well Control for Completions and Interventions, Gulf Professional Publishing, 2018.

[34] D.J. Lary, A.H. Alavi, A.H. Gandomi, A.L. Walker, Machine learning in geosciences and remote sensing, Geoscience Frontiers 7 (1) (2016) 3–10.

[35] E. Allison, B. Mandler, Subsurface data in the oil and gas industry, https://www.americangeosciences.org/geoscience-currents/subsurface-data-oil-and-gas-industry, 2018. (Accessed 23 July 2018).

[36] Big data growth continues in seismic surveys, https://bit.ly/2BVLsOs, 2015. (Accessed 2 September 2015).

[37] Oil and gas well data, https://www.kgslibrary.com/.

[38] D. Liu, G. Zhu, J. Zhang, K. Huang, Wireless data acquisition for edge learning: Importance-aware retransmission, in: 2019 IEEE 20th International Workshop on Signal Processing Advances in Wireless Communications (SPAWC), IEEE, 2019, pp. 1–5.

[39] R. Michel, Predictive analytics hit the midstream, https://www.oilandgaseng.com/articles/predictive-analytics-hit-the-midstream/. (Accessed 25 January 2022).

[40] X. Shi, G. Liu, X. Gong, J. Zhang, J. Wang, H. Zhang, An efficient approach for real-time prediction of rate of penetration in offshore drilling, Mathematical Problems in Engineering 2016 (2016).

[41] A. Takbiri-Borujeni, E. Fathi, T. Sun, R. Rahmani, M. Khazaeli, et al., Drilling performance monitoring and optimization: a data-driven approach, Journal of Petroleum Exploration and Production Technology 9 (4) (2019) 2747–2756.

[42] M.A. Ahmadi, S.R. Shadizadeh, K. Shah, A. Bahadori, An accurate model to predict drilling fluid density at wellbore conditions, Egyptian Journal of Petroleum 27 (1) (2018) 1–10.

[43] A.K. Abbas, A.A. Bashikh, H. Abbas, H.Q. Mohammed, Intelligent decisions to stop or mitigate lost circulation based on machine learning, Energy 183 (2019) 1104–1113.

[44] Petroleum Experts donate commercial software, https://www.cardiff.ac.uk/news/view/1483738-petroleum-experts-donate-commercial-software/. (Cited: April 24, 2019).

[45] C.P. Ervin, Theory of the Bouguer anomaly, Geophysics 42 (7) (1977) 1468.

[46] N. Sarapulov, R. Khabibullin, et al., Application of big data tools for unstructured data analysis to improve ESP operation efficiency (Russian), SPE Russian Petroleum Technology Conference, Society of Petroleum Engineers, 2017.

[47] S. Gupta, L. Saputelli, M. Nikolaou, et al., Big data analytics workflow to safeguard ESP operations in real-time, in: SPE North America Artificial Lift Conference and Exhibition, Society of Petroleum Engineers, 2016.

[48] M. Abhinesh, L. Archana, G. Ashwini, N.J. Deeptha, S.S. Gayathri, R. HariPriya, Descriptive analysis of oil and gas industries, International Research Journal of Engineering and Technology (Apr. 2019).

[49] S.N. Kayum, M. Rogowski, High-performance computing applications' transition to the cloud in the oil & gas industry, in: Proceedings of IEEE High Performance Extreme Computing Conference (HPEC), 2019.

[50] F. Li, D. Wang, F. Yan, F. Song, ElasticBroker: Combining HPC with cloud to provide realtime insights into simulations, arXiv preprint, arXiv:2010.04828, 2020.

[51] Z. Ndamase, The impact of data governance on corporate performance: the case of a petroleum company, Master's thesis, University of Cape Town, 2014.

[52] Scada systems: Improving efficiency in the oil and gas industry, https://blog.sitepro.com/resources/blog/scada-systems-improving-efficiency-in-oil-and-gas/, 2018. (Accessed 27 July 2018).

[53] IoT revolution in the oil and gas industry, https://www.digiteum.com/iot-oil-gas-industry, 2019. (Accessed 29 October 2019).

[54] M. Faber, K. vd Zwaag, H. Rocha, E. Pereira, M. Segatto, J. Silva, Performance evaluation of LoRaWAN applied to smart monitoring in onshore oil industries, in: 2019 SBMO/IEEE MTT-S International Microwave and Optoelectronics Conference (IMOC), IEEE, 2019, pp. 1–3.

[55] L.N. Chavala, R. Singh, A. Gehlot, Performance evaluation of LoRa based sensor node and gateway architecture for oil pipeline management, International Journal of Electrical and Computer Engineering 12 (1) (2022) 974.

[56] IoT solutions for upstream oil and gas, https://www.intel.com/content/dam/www/public/us/en/documents/solution-briefs/iot-upstream-oil-gas-solution-brief.pdf, 2016. (Accessed 29 October 2019).

[57] A. Khodabakhsh, I. Ari, M. Bakir, Cloud-based fault detection and classification for oil & gas industry, arXiv preprint, arXiv:1705.04583, 2017.

[58] R. Thavarajah, Y. Agbor, D. Tishechkin, Predicting the failure of turbofan engines using SpeedWise machine learning | Amazon Web Services, Mar. 23, 2022. (Online; accessed 17 April 2022).

[59] Predictive maintenance using machine learning | Implementations | AWS Solutions. (Online; accessed 17 April 2022).

[60] E.M. Kermani, V. Kapoor, S. Adeshina, J. Siri, Github – awslabs/predictive-maintenance-using-machine-learning: Set up end-to-end demo architecture for predictive maintenance issues with machine learning using Amazon SageMaker, May 18, 2021. (Online; accessed 17 April 2022).

[61] C.W. Remeljej, A.F.A. Hoadley, An exergy analysis of small-scale liquefied natural gas (LNG) liquefaction processes, Energy 31 (12) (2006) 2005–2019.

[62] C.S. Hsu, P.R. Robinson, Midstream Transportation, Storage, and Processing, Springer International Publishing, 2019, pp. 385–394.

[63] Oil transport, https://www.studentenergy.org/topics/ff-transport, 2019. (Accessed 1 July 2019).

[64] F. Post, How North America's oil production push has provided big growth for midstream firms, https://financialpost.com/commodities/energy/how-north-americas-oil-production-push-has-provided-big-growth-for-midstream-firms. (Accessed 16 February 2022).

[65] J. Hu, P. Bhowmick, F. Arvin, A. Lanzon, B. Lennox, Cooperative control of heterogeneous connected vehicle platoons: An adaptive leader-following approach, IEEE Robotics and Automation Letters 5 (2) (2020) 977–984.

[66] A. Sarker, C. Qiu, H. Shen, Connectivity maintenance for next-generation decentralized vehicle platoon networks, IEEE/ACM Transactions on Networking (2020).

[67] S.K. Dwivedi, R. Amin, S. Vollala, R. Chaudhry, Blockchain-based secured event-information sharing protocol in internet of vehicles for smart cities, Computers & Electrical Engineering 86 (2020) 106719.

[68] Y. Fu, F.R. Yu, C. Li, T.H. Luan, Y. Zhang, Vehicular blockchain-based collective learning for connected and autonomous vehicles, IEEE Wireless Communications 27 (2) (2020) 197–203.

[69] Route optimization helped a petroleum company to streamline multi-stop route planning | A Quantzig success story | Business Wire, May 15, 2020. (Online; accessed 1 May 2022).

[70] Oil and gas | Oil and gas industry transportation route optimization, Sep. 22, 2016. (Online; accessed 1 May 2022).

[71] P. Nimmanonda, V. Uraikul, C.W. Chan, P. Tontiwachwuthikul, Computer-aided simulation model for natural gas pipeline network system operations, Journal of Industrial & Engineering Chemistry Research 43 (4) (2004) 990–1002.

[72] O.F.M. El-Mahdy, M.E.H. Ahmed, S. Metwalli, Computer aided optimization of natural gas pipe networks using genetic algorithm, Applied Soft Computing 10 (4) (2010) 1141–1150.

[73] D. Shauers, Oil and gas pipeline design, market continues evolving, https://www.oilandgaseng.com/articles/oil-and-gas-pipeline-design-market-continues-evolving/, 2019. (Accessed 3 June 2019).

[74] B. Shiklo, Make data the new oil: IIoT-enabled predictive maintenance for oil and gas, https://connectedworld.com/make-data-the-new-oil-iiot-enabled-predictive-maintenance-for-the-oil-and-gas-industry/, 2019. (Accessed 12 February 2019).

[75] Sensors for oil and gas pipeline monitoring, https://www.prnewswire.com/news-releases/sensors-for-oil-and-gas-pipeline-monitoring-300445608.html, 2017. (Accessed 25 April 2017).

[76] G. Kabir, R. Sadiq, S. Tesfamariam, A fuzzy Bayesian belief network for safety assessment of oil and gas pipelines, Structure and Infrastructure Engineering 12 (8) (2016) 874–889.

[77] Pipeline leak detection for oil and gas using IoT, https://www.biz4intellia.com/blog/pipeline-leak-detection-with-iot-in-oil-and-gas/, 2019. (Accessed 22 April 2019).

[78] G. He, Y. Li, Y. Huang, L. Sun, K. Liao, A framework of smart pipeline system and its application on multiproduct pipeline leakage handling, Energy 188 (2019) 116031.

[79] How IoT is transforming oil & gas pipeline management, https://www.iotforall. com/how-iot-is-transforming-oil-gas-pipeline-management/, 2019. (Accessed 22 May 2019).

[80] N. Dutta, S. Umashankar, V.A. Shankar, S. Padmanaban, Z. Leonowicz, P. Wheeler, Centrifugal pump cavitation detection using machine learning algorithm technique, in: 2018 IEEE International Conference on Environment and Electrical Engineering and 2018 IEEE Industrial and Commercial Power Systems Europe (EEEIC/I&CPS Europe), 2018, pp. 1–6.

[81] T. Barbariol, E. Feltresi, G.A. Susto, Machine learning approaches for anomaly detection in multiphase flow meters, IFAC-PapersOnLine 52 (11) (2019) 212–217.

[82] M.A.A.-O. Salah, K. Hui, L. Hee, M.S. Leong, A.A.-H. Mahdi, A.M. Abdelrhman, Y. Ali, Automated valve fault detection based on acoustic emission parameters and artificial neural network, in: MATEC Web of Conferences, vol. 255, 2019, 02013.

[83] M. Nasiri, M. Mahjoob, H. Vahid-Alizadeh, Vibration signature analysis for detecting cavitation in centrifugal pumps using neural networks, in: 2011 IEEE International Conference on Mechatronics, IEEE, 2011, pp. 632–635.

[84] Rapidly detect pipeline leaks with cloud-based solution that offers accurate alarming, https://www.zedisolutions.com/hubfs/Application%20Notes/application-note-rapidly-detect-pipeline-leaks-cloud-based-solution-that-offers-accurate-alarming-en-7239264.pdf.

[85] Types of storage tanks, https://www.pipingengineer.org/types-of-storage-tanks/, 2015.

[86] Oil storage tanks with a floating roof, https://leakwise.com/applications/oil-storage-tanks-with-a-floating-roof/.

[87] T. Wang, Y. Li, S. Yu, Y. Liu, Estimating the volume of oil tanks based on high-resolution remote sensing images, Remote Sensing 11 (7) (2019) 793.

[88] Y. Yu, L. Zhou, Acoustic emission signal classification based on support vector machine, TELKOMNIKA Indonesian Journal of Electrical Engineering 10 (5) (2012) 1027–1032.

[89] J.B. Macêdo, A machine learning-based methodology for automated classification of risks in an oil refinery, Master's thesis, Universidade Federal de Pernambuco, 2019.

[90] R.K. Lattanzio, Methane and Other Air Pollution Issues in Natural Gas Systems, Congressional Research Service, 2017.

[91] J. Ni, H. Yang, J. Yao, Z. Li, P. Qin, Toxic gas dispersion prediction for point source emission using deep learning method, Human and Ecological Risk Assessment: An International Journal 26 (2) (2020) 557–570.

[92] L. Klein, M. Ramachandran, T. van Kessel, D. Nair, N. Hinds, H. Hamann, N. Sosa, Wireless sensor networks for fugitive methane emissions monitoring in oil and gas industry, in: 2018 IEEE International Congress on Internet of Things (ICIOT), IEEE, 2018, pp. 41–48.

[93] Best management practices for flare reduction in flare stacks, https://www. aspireenergy.com/best-management-practices-for-flare-reduction-in-flare-stacks/, 2019. (Accessed 29 October 2019).

[94] Flaring and venting, http://www.capp.ca/environmentCommunity/ airClimateChange/Pages/FlaringVenting.aspx, 2014. (Accessed 5 April 2014).

[95] Natural gas and global gas flaring reduction, http://go.worldbank.org/ 7WIFCC42A0, 2014. (Accessed 5 April 2014).

[96] US oil producers turn to crypto mining to reduce flaring, https://bit.ly/3KdQGU0, 2022. (Accessed 24 March 2022).

[97] Exxon weighs taking gas-to-bitcoin pilot to four countries, https:// www.bnnbloomberg.ca/exxon-weighs-taking-gas-to-bitcoin-pilot-to-four-countries-1.1742495, 2022. (Accessed 24 March 2022).

[98] C.H. Peterson, S.D. Rice, J.W. Short, D. Esler, J.L. Bodkin, B.E. Ballachey, D.B. Irons, Long-term ecosystem response to the Exxon Valdez oil spill, Science 302 (5653) (2003) 2082–2086.

[99] Z. Jiao, G. Jia, Y. Cai, A new approach to oil spill detection that combines deep learning with unmanned aerial vehicles, Computers & Industrial Engineering 135 (2019) 1300–1311.

[100] T.K. Fataliyev, S.A. Mehdiyev, Analysis and new approaches to the solution of problems of operation of oil and gas complex as cyber-physical system, International Journal of Information Technology and Computer Science 10 (11) (2018) 67–76.

[101] D.E. Denning, Stuxnet: What has changed?, Future Internet 4 (3) (2012) 672–687.

[102] O. Griffin, Colombia ELN guerrillas claim responsibility for attacks on oilinfrastructure | Reuters, Oct. 15, 2021. (Online; accessed 25 April 2022).

[103] F. Steinhäusler, P. Furthner, W. Heidegger, S. Rydell, L. Zaitseva, Security risks to the oil and gas industry: Terrorist capabilities strategic insights VII (1) (February 2008).

[104] S. Brueske, C. Kramer, A. Fisher, Bandwidth study on energy use and potential energy saving opportunities in US pulp and paper manufacturing, Tech. rep., 2015.

[105] Q. Wu, L. Wang, Z. Zhu, Research of pre-stack AVO elastic parameter inversion problem based on hybrid genetic algorithm, Cluster Computing 20 (4) (2017) 3173–3183.

[106] M.A. Durrani, A. Avila, J. Rafael, I. Ahmad, An integrated mechanism of genetic algorithm and Taguchi method for cut-point temperatures optimization of crude distillation unit, in: 2018 International Conference on Computing, Mathematics and Engineering Technologies (iCoMET), IEEE, 2018, pp. 1–6.

[107] B. McMahan, E. Moore, D. Ramage, S. Hampson, B.A. y Arcas, Communication-efficient learning of deep networks from decentralized data, in: Artificial Intelligence and Statistics, 2017, pp. 1273–1282.

[108] V. Veillon, C. Denninnart, M.A. Salehi, F-FDN: Federation of fog computing systems for low latency video streaming, in: Proceedings of the 3rd IEEE International Conference on Fog and Edge Computing, 2019, pp. 1–9.

[109] L. Matijašević, A. Vučković, I. Dejanović, Analysis of cooling water systems in a petroleum refinery, Chemical and Biochemical Engineering Quarterly 28 (4) (2014) 451–457.

[110] Cooling towers | Emerson US. (Online; accessed 23 April 2022).

[111] N.K. Verma, D. Kumar, I. Kumar, A. Ashok, Automation of boiler process at thermal power plant using sensors and IoT, Journal of Statistics and Management Systems 21 (4) (2018) 675–683.

[112] K. Lindqvist, Z.T. Wilson, E. Næss, N.V. Sahinidis, A machine learning approach to correlation development applied to fin-tube bundle heat exchangers, Energies 11 (12) (2018) 3450.

[113] J.H. Gary, Petroleum refining, in: Encyclopedia of Physical Science and Technology, third edition, 2003, pp. 741–761.

[114] H. Sildir, Y. Arkun, U. Canan, S. Celebi, U. Karani, I. Er, Dynamic modeling and optimization of an industrial fluid catalytic cracker, Journal of Process Control 31 (2015) 30–44.

[115] L. Eriksen, The impact of digital on unplanned downtime, 2016.

[116] Blockchain for oil and gas – IBM Blockchain | IBM. (Online; accessed 28 April 2022).

[117] V. Mullineaux, Collaboration: Key to the first multicloud blockchain deployment IBM Supply Chain and Blockchain Blog, Dec. 2, 2019. (Online; accessed 28 April 2022).

[118] E. Denaro, Vertrax revolutionizes their supply chain logistics using IBM Blockchain on multicloud – Chateaux, Dec. 28, 2021. (Online; accessed 28 April 2022).

[119] S. Hashemi, D. Mowla, F. Esmaeilzadeh, Assessment and simulation of gaseous dispersion by computational fluid dynamics (CFD): A case study of Shiraz oil refining company, American Journal of Environmental Science and Engineering 4 (2) (2020) 17.

[120] A. Rahimi, T. Tavakoli, S. Zahiri, Computational fluid dynamics (CFD) modeling of gaseous pollutants dispersion in low wind speed condition: Isfahan refinery, a case study, Petroleum Science and Technology 32 (11) (2014) 1318–1326.

[121] Oil & gas: Revamping upstream, midstream, and downstream segment with IIoT. (Online; accessed 19 April 2022).

[122] I. Kilovaty, Cybersecuring the pipeline, Houston Law Review 60 (2023).

[123] Cyberattack causes chaos at key European oil terminals | S&P Global Commodity Insights, Feb. 3, 2022. (Online; accessed 24 April 2022).

[124] Some of the biggest cyberattacks on energy infrastructure in recent years – Prospero Events Group, Mar. 24, 2022. (Online; accessed 24 April 2022).

[125] IT-OT convergence in oil & gas: Top strategies & benefits. (Online; accessed 24 April 2022).

[126] Q. Zhang, Z. Qian, S. Wang, L. Yuan, B. Gong, Productivity drain or productivity gain? The effect of new technology adoption in the oilfield market, Energy Economics 108 (2022) 105930.

[127] Understand the role of IT/OT convergence in oil and gas. (Online; accessed 26 April 2022).

[128] E.D. Knapp, J. Langill, Industrial Network Security: Securing Critical Infrastructure Networks for Smart Grid, SCADA, and Other Industrial Control Systems, Syngress, 2014.

[129] N. Jan, A. Nasir, M.S. Alhilal, S.U. Khan, D. Pamucar, A. Alothaim, Investigation of cyber-security and cyber-crimes in oil and gas sectors using the innovative structures of complex intuitionistic fuzzy relations, Entropy 23 (9) (2021) 1112.

[130] S. Mandal, A. Chirputkar, Digital transformation and cyber disruption in the oil and gas industry, PalArch's Journal of Archaeology of Egypt/Egyptology 17 (6) (2020) 4402–4416.

[131] F. Hacquebord, C. Pernet, Drilling deep: A look at cyberattacks on the oil and gas industry, https://www.trendmicro.com/vinfo/it/security/news/internet-of-things/drilling-deep-a-look-at-cyberattacks-on-the-oil-and-gas-industry, 2019.

[132] R.S. Kusuma, R. Umar, I. Riadi, Network forensics against Ryuk ransomware using trigger, acquire, analysis, report, and action (TAARA) method, in: Kinetik: Game Technology, Information System, Computer Network, Computing, Electronics, and Control, 2021.

[133] B. Conde Gallego, J. Drexl, IoT connectivity standards: how adaptive is the current SEP regulatory framework?, IIC-International Review of Intellectual Property and Competition Law 50 (1) (2019) 135–156.

[134] M. Zborowski, As oil and gas data multiply, so do the cybersecurity threats, Sep. 20, 2019. (Online; accessed 30 April 2022).

[135] K. Rick, K. Iyer, Countering the threat of cyberattacks in oil and gas, Jan. 8, 2021. (Online; accessed 30 April 2022).

[136] L. O'Brien, Data diodes offer cost-effective way | ARC Advisory. (Online; accessed 30 April 2022).

[137] M.T. Hossain, S. Badsha, H. Shen, Porch: A novel consensus mechanism for blockchain-enabled future SCADA systems in smart grids and industry 4.0, in: 2020 IEEE International IOT, Electronics and Mechatronics Conference (IEMTRONICS), IEEE, 2020, pp. 1–7.

[138] D. Ongaro, J. Ousterhout, In search of an understandable consensus algorithm, in: 2014 USENIX Annual Technical Conference (Usenix ATC 14), 2014, pp. 305–319.

[139] M.B. Mollah, J. Zhao, D. Niyato, K.-Y. Lam, X. Zhang, A.M. Ghias, L.H. Koh, L. Yang, Blockchain for future smart grid: A comprehensive survey, IEEE Internet of Things Journal 8 (1) (2020) 18–43.

[140] R. Zhang, R. Xue, L. Liu, Security and privacy on blockchain, ACM Computing Surveys 52 (3) (2019) 1–34.

[141] R. Di Pietro, X. Salleras, M. Signorini, E. Waisbard, A blockchain-based trust system for the internet of things, in: Proceedings of the 23rd ACM Symposium on Access Control Models and Technologies, 2018, pp. 77–83.

[142] S. Abdul-Jabbar, Council post: Where blockchain technology can disrupt the oil and gas industry, Sep. 27, 2021. (Online; accessed 29 April 2022).

[143] Y. Wang, A. Kogan, Designing confidentiality-preserving blockchain-based transaction processing systems, International Journal of Accounting Information Systems 30 (2018) 1–18.

[144] B. Körbel, M. Sigwart, P. Frauenthaler, M. Sober, S. Schulte, Blockchain-based result verification for computation offloading, in: International Conference on Service-Oriented Computing, Springer, 2021, pp. 99–115.

[145] J. Grundy, Human-centric software engineering for next generation cloud-and edge-based smart living applications, in: 2020 20th IEEE/ACM International Symposium on Cluster, Cloud and Internet Computing (CCGRID), IEEE, 2020, pp. 1–10.

[146] J. Grundy, We need to get smart(er) in software design, May 25, 2020. (Online; accessed 27 April 2022).

[147] S. Lohr, Facial recognition is accurate, if you're a white guy, in: Ethics of Data and Analytics, Auerbach Publications, 2018, pp. 143–147.

[148] M. Madden, A. Lenhart, S. Cortesi, U. Gasser, M. Duggan, A. Smith, M. Beaton, Teens, social media, and privacy, Pew Research Center 21 (1055) (2013) 2–86.

[149] T. Bolukbasi, K.-W. Chang, J.Y. Zou, V. Saligrama, A.T. Kalai, Man is to computer programmer as woman is to homemaker? Debiasing word embeddings, Advances in Neural Information Processing Systems 29 (2016).

[150] D. Wesley, L.A. Dau, Complacency and automation bias in the Enbridge pipeline disaster, Ergonomics in Design 25 (1) (2017) 17–22.

[151] Automation bias, https://databricks.com/glossary/automation-bias.

[152] K. Rick, I. Martén, U. Von Lonski, Untapped Reserves, Promoting Gender Balance in Oil and Gas, World Petroleum Council, Boston Consulting Group, 2017.

[153] 'Double jeopardy' prevents culturally diverse women from top leadership roles | HRD Australia. (Online; accessed 30 April 2022).

[154] R.W. McLeod, The impact of styles of thinking and cognitive bias on how people assess risk and make real-world decisions in oil and gas operations, Oil and Gas Facilities 5 (05) (2016).

[155] A. Tversky, D. Kahneman, Judgment under uncertainty: Heuristics and biases: Biases in judgments reveal some heuristics of thinking under uncertainty, Science 185 (4157) (1974) 1124–1131.

[156] M. Prause, J. Weigand, The rig: A leadership practice game to train on debiasing techniques, in: 2017 Winter Simulation Conference (WSC), IEEE, 2017, pp. 4312–4323.

[157] D. Mathieson, et al., Forces that will shape intelligent-wells development, Journal of Petroleum Technology 59 (Aug. 2007) 14–16.

[158] H.K. White, P.-Y. Hsing, W. Cho, T.M. Shank, E.E. Cordes, A.M. Quattrini, R.K. Nelson, R. Camilli, A.W.J. Demopoulos, C.R. German, J.M. Brooks, H.H. Roberts, W. Shedd, C.M. Reddy, C.R. Fisher, Impact of the deepwater horizon oil spill on a deep-water coral community in the Gulf of Mexico, Proceedings of the National Academy of Sciences 109 (Feb. 2012) 20303–20308.

[159] Cisco, A New Reality for Oil & Gas: Data Management and Analytics, White paper, April 2015.

[160] S. Prabhu, E. Gajendran, N. Balakumar, Smart oil field management using wireless communication techniques, International Journal of Inventions in Engineering & Science Technology (Jan. 2016) 2454–9584.

[161] N.G. Franconi, A.P. Bunger, E. Sejdić, M.H. Mickle, Wireless communication in oil and gas wells, Energy Technology 2 (12) (Oct. 2014) 996–1005.

[162] WoodMackenzie, Why are some deepwater plays still attractive?, White paper, Sep. 2017.

[163] ABB, Field Area Communication Networks for Digital Oil and Gas Fields, White paper, October 2014.

[164] M.A. Hayes, M.A.M. Capretz, Contextual anomaly detection in big sensor data, in: Proceedings of IEEE International Congress on Big Data, June 2014, pp. 64–71.

[165] David Riddle, Danger and detection of hydrogen sulphide gas in oil and gas exploration and production, White paper, April 2009.

[166] A.A. Bisu, A. Purvis, K. Brigham, H. Sun, A framework for end-to-end latency measurements in a satellite network environment, in: 2018 IEEE International Conference on Communications (ICC), May 2018, pp. 1–6.

[167] J. Pickering, S. Sengupta, M. Pfitzinger, et al., Adopting cloud technology to enhance the digital oilfield, in: Proceedings of International Petroleum Technology Conference, Dec. 2015.

[168] J. Cho, G. Lim, T. Biobaku, S. Kim, H. Parsaei, Safety and security management with unmanned aerial vehicle (UAV) in oil and gas industry, Journal of Procedia Manufacturing 3 (Jul. 2015) 1343–1349.

[169] B.I. Ismail, E.M. Goortani, M.B. Ab Karim, W.M. Tat, S. Setapa, J.Y. Luke, O.H. Hoe, Evaluation of docker as edge computing platform, in: Proceedings of the IEEE Conference on Open Systems (ICOS), Aug. 2015, pp. 130–135.

[170] W. Hu, Y. Gao, K. Ha, J. Wang, B. Amos, Z. Chen, P. Pillai, M. Satyanarayanan, Quantifying the impact of edge computing on mobile applications, in: Proceedings of the 7th ACM SIGOPS Asia-Pacific Workshop on Systems, APSys '16, 2016, pp. 5:1–5:8.

[171] H.-W. Lee, M.-I. Roh, Review of the multibody dynamics in the applications of ships and offshore structures, Journal of Ocean Engineering 167 (2018) 65–76.

[172] R.B. Rodrigo Escobar, David Akopian, A sensor data format incorporating battery charge information for smartphone-based mHealth applications, vol. 9411, Mar. 2015.

[173] M. Skedsmo, R. Ayasse, N. Soleng, M. Indregard, et al., Oil spill detection and response using satellite imagery, insight to technology and regulatory context, in: Proceedings of the Society of Petroleum Engineers (SPE) International Conference and Exhibition on Health, Safety, Security, Environment, and Social Responsibility, Apr. 2016.

[174] W.Z. Khan, M.Y. Aalsalem, M.K. Khan, M.S. Hossain, M. Atiquzzaman, A reliable internet of things based architecture for oil and gas industry, in: Proceedings of 19th International Conference on Advanced Communication Technology, Feb. 2017, pp. 705–710.

[175] X. Li, M.A. Salehi, M. Bayoumi, R. Buyya, CVSS: A cost-efficient and QoS-aware video streaming using cloud services, in: Proceedings of 16th IEEE/ACM International Symposium on Cluster, Cloud and Grid Computing (CCGrid), CCGRID '16, May 2016, pp. 106–115.

[176] C. Denninnart, J. Gentry, M.A. Salehi, Improving robustness of heterogeneous serverless computing systems via probabilistic task pruning, in: Proceedings of the 33rd IEEE International Parallel & Distributed Processing Symposium Workshops, May. 2019.

[177] C.O. Diaz, M. Guzek, J.E. Pecero, G. Danoy, P. Bouvry, S.U. Khan, Energy-aware fast scheduling heuristics in heterogeneous computing systems, in: Proceedings of International Conference on High Performance Computing & Simulation, Jul. 2011, pp. 478–484.

[178] C. Sonmez, A. Ozgovde, C. Ersoy, EdgeCloudSim: An environment for performance evaluation of edge computing systems, in: Proceedings of the 2nd International Conference on Fog and Mobile Edge Computing, FMEC '17, May 2017, pp. 39–44.

[179] M.A. Salehi, J. Smith, A.A. Maciejewski, H.J. Siegel, E.K. Chong, J. Apodaca, L.D. Briceño, T. Renner, V. Shestak, J. Ladd, et al., Stochastic-based robust dynamic resource allocation for independent tasks in a heterogeneous computing system, Journal of Parallel and Distributed Computing 97 (Jun. 2016) 96–111.

[180] R.F. Hussain, M.A. Salehi, A. Kovalenko, S. Salehi, O. Semiari, Robust resource allocation using edge computing for smart oil fields, in: Proceedings of the 24th International Conference on Parallel and Distributed Processing Techniques & Applications, Aug. 2018.

[181] M.N. Jha, J. Levy, Y. Gao, Advances in remote sensing for oil spill disaster management: state-of-the-art sensors technology for oil spill surveillance, Sensors 8 (1) (2008) 236–255.

[182] A. Vanjare, C. Arvind, S. Omkar, J. Kishore, V. Kumar, GEP algorithm for oil spill detection and differentiation from lookalikes in RISAT SAR images, in: Soft Computing for Problem Solving, Springer, 2019, pp. 435–446.

[183] S. Tong, X. Liu, Q. Chen, Z. Zhang, G. Xie, Multi-feature based ocean oil spill detection for polarimetric SAR data using random forest and the self-similarity parameter, Remote Sensing 11 (4) (2019) 451.

[184] G. Orfanidis, K. Ioannidis, K. Avgerinakis, S. Vrochidis, I. Kompatsiaris, A deep neural network for oil spill semantic segmentation in SAR images, in: 2018 25th IEEE International Conference on Image Processing (ICIP), IEEE, 2018, pp. 3773–3777.

[185] M.S. Alam, P. Sidike, Trends in oil spill detection via hyperspectral imaging, in: 2012 7th International Conference on Electrical and Computer Engineering, IEEE, 2012, pp. 858–862.

[186] M. Krestenitis, G. Orfanidis, K. Ioannidis, K. Avgerinakis, S. Vrochidis, I. Kompatsiaris, Oil spill identification from satellite images using deep neural networks, Remote Sensing 11 (15) (2019) 1762.

[187] A.H.S. Solberg, Remote sensing of ocean oil-spill pollution, Proceedings of the IEEE 100 (10) (2012) 2931–2945.

[188] M.F. Fingas, C.E. Brown, Review of oil spill remote sensing, Spill Science & Technology Bulletin 4 (4) (1997) 199–208.

[189] A.S. Solberg, G. Storvik, R. Solberg, E. Volden, Automatic detection of oil spills in ERS SAR images, IEEE Transactions on Geoscience and Remote Sensing 37 (4) (1999) 1916–1924.

[190] B. Fiscella, A. Giancaspro, F. Nirchio, P. Pavese, P. Trivero, Oil spill detection using marine SAR images, International Journal of Remote Sensing 21 (18) (2000) 3561–3566.

[191] H. Espedal, Satellite SAR oil spill detection using wind history information, International Journal of Remote Sensing 20 (1) (1999) 49–65.

[192] F. Del Frate, A. Petrocchi, J. Lichtenegger, G. Calabresi, Neural networks for oil spill detection using ERS-SAR data, IEEE Transactions on Geoscience and Remote Sensing 38 (5) (2000) 2282–2287.

[193] D.L. de Souza, A.D.D. Neto, W. da Mata, Intelligent system for feature extraction of oil slick in SAR images: Speckle filter analysis, in: Neural Information Processing, Springer, Berlin, Heidelberg, 2006, pp. 729–736.

[194] E.E. Konstantinidou, P. Kolokoussis, K. Topouzelis, I. Moutzouris-Sidiris, An open source approach for oil spill detection using Sentinel-1 SAR images, in: Seventh International Conference on Remote Sensing and Geoinformation of the Environment (RSCy2019), in: Proc. SPIE, vol. 11174, International Society for Optics and Photonics, 2019, pp. 332–337.

[195] O. Garcia-Pineda, B. Zimmer, M. Howard, W. Pichel, X. Li, I.R. MacDonald, Using SAR images to delineate ocean oil slicks with a texture-classifying neural network algorithm (TCNNA), Canadian Journal of Remote Sensing 35 (5) (2009) 411–421.

[196] X. Yu, H. Zhang, C. Luo, H. Qi, P. Ren, Oil spill segmentation via adversarial f-divergence learning, IEEE Transactions on Geoscience and Remote Sensing 56 (Sep. 2018) 4973–4988.

[197] A.-J. Gallego, P. Gil, A. Pertusa, R.B. Fisher, Semantic segmentation of SLAR imagery with convolutional LSTM selectional autoencoders, Remote Sensing 11 (12) (2019) 1402.

[198] J. Long, E. Shelhamer, T. Darrell, Fully convolutional networks for semantic segmentation, in: Proceedings of the IEEE Conference on Computer Vision and Pattern Recognition, 2015, pp. 3431–3440.

[199] M. ul Hassan, VGG16 – convolutional network for classification and detection, https://neurohive.io/en/popular-networks/vgg16/, Nov. 20, 2018.

[200] K. Simonyan, A. Zisserman, Very deep convolutional networks for large-scale image recognition, arXiv preprint, arXiv:1409.1556, 2014.

[201] N. Shibuya, Up-sampling with transposed convolution, https://medium.com/activating-robotic-minds/up-sampling-with-transposed-convolution-9ae4f2df52d0, Nov. 13, 2017.

[202] F. Chollet, et al., Keras, https://github.com/fchollet/keras, 2015.

[203] J. Jordan, Evaluating image segmentation models, https://www.jeremyjordan.me/evaluating-image-segmentation-models/, May 30, 2018.

[204] H. Lu, L. Guo, M. Azimi, K. Huang, Oil and gas 4.0 era: A systematic review and outlook, Journal of Computers in Industry 111 (2019) 68–90.

[205] X. Li, M.A. Salehi, M. Bayoumi, N.-F. Tzeng, R. Buyya, Cost-efficient and robust on-demand video transcoding using heterogeneous cloud services, IEEE Transactions on Parallel and Distributed Systems 29 (3) (2017) 556–571.

[206] I. Dincer, C. Acar, A review on clean energy solutions for better sustainability, International Journal of Energy Research 39 (5) (2015) 585–606.

[207] A.J. Dunnings, T.P. Breckon, Experimentally defined convolutional neural network architecture variants for non-temporal real-time fire detection, in: Proceedings of 25th IEEE International Conference on Image Processing (ICIP), IEEE, 2018, pp. 1558–1562.

[208] F. Aliyu, T. Sheltami, Development of an energy-harvesting toxic and combustible gas sensor for oil and gas industries, Journal of Sensors and Actuators B: Chemical 231 (2016) 265–275.

[209] A.E. Eshratifar, M.S. Abrishami, M. Pedram, JointDNN: an efficient training and inference engine for intelligent mobile cloud computing services, Journal of IEEE Transactions on Mobile Computing (2019).

[210] Q. Zhang, M. Zhang, M. Wang, W. Sui, C. Meng, J. Yang, W. Kong, X. Cui, W. Lin, Efficient deep learning inference based on model compression, in: Proceedings of the IEEE Conference on Computer Vision and Pattern Recognition Workshops, 2018, pp. 1695–1702.

[211] Z. Han, H. Tan, X.-Y. Li, S.H.-C. Jiang, Y. Li, F.C. Lau, OnDisc: Online latency-sensitive job dispatching and scheduling in heterogeneous edge-clouds, IEEE/ACM Transactions on Networking 27 (6) (2019) 2472–2485.

[212] I. Bermudez, S. Traverso, M. Mellia, M. Munafo, Exploring the cloud from passive measurements: The Amazon AWS case, in: Proceedings of IEEE INFOCOM, 2013, pp. 230–234.

[213] A. Luckow, M. Cook, N. Ashcraft, E. Weill, E. Djerekarov, B. Vorster, Deep learning in the automotive industry: Applications and tools, in: Proceedings of International Conference on Big Data (Big Data), 2016, pp. 3759–3768.

[214] A. Mokhtari, C. Denninnart, M.A. Salehi, Autonomous task dropping mechanism to achieve robustness in heterogeneous computing systems, in: Proceedings of the 29th Heterogeneity in Computing Workshop (HCW 2019), May 2020.

[215] J. Gentry, C. Denninnart, M. Amini Salehi, Robust dynamic resource allocation via probabilistic task pruning in heterogeneous computing systems, in: Proceedings of the 33rd IEEE International Parallel & Distributed Processing Symposium, IPDPS '19, May 2019.

[216] C. Xia, J. Zhao, H. Cui, X. Feng, J. Xue, DNNTune: Automatic benchmarking DNN models for mobile-cloud computing, ACM Transactions on Architecture and Code Optimization 16 (4) (2019).

[217] K. Keahey, P. Riteau, D. Stanzione, T. Cockerill, J. Mambretti, P. Rad, P. Ruth, Chameleon: a scalable production testbed for computer science research, in: Contemporary High Performance Computing: From Petascale Toward Exascale, vol. 3, May 2019, pp. 123–148, Ch. 5.

[218] T. Nguyen, R.G. Gosine, P. Warrian, A systematic review of big data analytics for oil and gas industry 4.0, IEEE Access 8 (2020) 61183–61201.

[219] M.Z. Alom, T.M. Taha, C. Yakopcic, S. Westberg, P. Sidike, M.S. Nasrin, B.C. Van Esesn, A.A.S. Awwal, V.K. Asari, The history began from AlexNet: A comprehensive survey on deep learning approaches, arXiv preprint, arXiv:1803.01164, 2018.

[220] D. Anguita, A. Ghio, L. Oneto, X. Parra, J.L. Reyes-Ortiz, A public domain dataset for human activity recognition using smartphones, in: ESANN, 2013.

[221] Z. Huang, C.O. Dumitru, Z. Pan, B. Lei, M. Datcu, Classification of large-scale high-resolution SAR images with deep transfer learning, Journal of Geoscience and Remote Sensing Letters (2020).

[222] A. Mustafa, M. Alfarraj, G. AlRegib, Estimation of acoustic impedance from seismic data using temporal convolutional network, arXiv preprint, arXiv:1906.02684, 2019.

[223] R. Versteeg, The Marmousi experience; velocity model determination on a synthetic complex data set, Journal of the Leading Edge 13 (1994) 927–936.

[224] A. Chenebert, T.P. Breckon, A. Gaszczak, A non-temporal texture driven approach to real-time fire detection, in: 2011 18th IEEE International Conference on Image Processing, IEEE, 2011, pp. 1741–1744.

[225] C.R. Steffens, R.N. Rodrigues, S.S. d. Costa Botelho, Non-stationary VFD evaluation kit: Dataset and metrics to fuel video-based fire detection development, in: Robotics, Springer, 2016, pp. 135–151.

[226] A. Brougois, M. Bourget, P. Lailly, M. Poulet, P. Ricarte, R. Versteeg, Marmousi, model and data, in: EAEG Workshop – Practical Aspects of Seismic Data Inversion, European Association of Geoscientists & Engineers, 1990, pp. 75–80.

[227] Marmousi case study, http://www.panoramatech.com/papers/book/bookse49.php. (Accessed 20 March 2022).

[228] T. Irons, Marmousi model, 2016.

[229] J. Varia, S. Mathew, et al., Overview of Amazon Web Services, Amazon Web Services, 2014, pp. 1–22.

[230] S.S. Ogden, T. Guo, Characterizing the deep neural networks inference performance of mobile applications, arXiv preprint, arXiv:1909.04783, 2019.

[231] H. Moradi, W. Wang, D. Zhu, Adaptive performance modeling and prediction of applications in multi-tenant clouds, in: 21st International Conference on High Performance Computing and Communications (HPCC), 2019, pp. 638–645.

[232] X. Li, M.A. Salehi, Y. Joshi, M.K. Darwich, B. Landreneau, M. Bayoumi, Perfor-
       mance analysis and modeling of video transcoding using heterogeneous cloud ser-
       vices, IEEE Transactions on Parallel and Distributed Systems 30 (4) (2019) 910–922.
[233] Z. Hanusz, J. Tarasinska, W. Zielinski, Shapiro–Wilk test with known mean, REVS-
       TAT Statistical Journal 14 (1) (2016).
[234] I.M. Chakravarty, J. Roy, R.G. Laha, Handbook of Methods of Applied Statistics,
       1967.
[235] D.J. Lilja, Measuring Computer Performance: A Practitioner's Guide, 2005.
[236] S. Patil, D.J. Lilja, Using resampling techniques to compute confidence intervals for
       the harmonic mean of rate-based performance metrics, IEEE Computer Architecture
       Letters 9 (1) (2010).
[237] B. Charyyev, A. Alhussen, H. Sapkota, E. Pouyoul, M.H. Gunes, E. Arslan, To-
       wards securing data transfers against silent data corruption, in: CCGRID, 2019,
       pp. 262–271.

Index

Printed in the United States
by Baker & Taylor Publisher Services